Effect of Emerging Processing Methods on the Food Quality

Shahin Roohinejad • Mohamed Koubaa
Ralf Greiner • Kumar Mallikarjunan
Editors

Effect of Emerging Processing Methods on the Food Quality

Advantages and Challenges

Springer

Editors
Shahin Roohinejad
Department of Food Science and Nutrition
University of Minnesota
St. Paul, MN, USA

Ralf Greiner
Department of Food Technology
and Bioprocess Engineering
Max Rubner-Institut
Federal Research Institute
of Nutrition and Food
Karlsruhe, Germany

Mohamed Koubaa
Ecole Supérieure de Chimie Organique
et Minérale
1 Allée du réseau Jean-Marie Buckmaster
Compiègne, France

Kumar Mallikarjunan
Department of Food Science and Nutrition
University of Minnesota
St. Paul, MN, USA

ISBN 978-3-030-18190-1 ISBN 978-3-030-18191-8 (eBook)
https://doi.org/10.1007/978-3-030-18191-8

This Springer imprint is published by the registered company Springer Nature Switzerland AG
The registered company address is: Gewerbestrasse 11, 6330 Cham, Switzerland

Preface

The food and beverage industries are widely searching for novel-innovative technologies to provide safe and healthy foods for the consumers. Although safe food products can be provided using thermal pasteurization and sterilization, heating the foods at high temperatures beyond a safety factor results in an unacceptable quality and nutrient retention. Thus, the application of alternative methods to minimize undesirable reactions such as thermal decomposition or degradation is highly desirable. Emerging processing technologies are promising methods to minimize heat-induced alterations in foods, and their applications have provided unprecedented opportunities for the food industry to make safe and high-quality health-promoting foodstuffs. These methods are useful not only for microorganisms and enzymes inactivation but also for improving the yield and development of ingredients and marketable foods with higher-quality and nutritional characteristics. Several studies have evaluated the effect of emerging processing methods such as high-pressure processing, pulsed electric fields, ultraviolet, pulsed light, irradiation, and ultrasounds on food quality. Many of these studies have reported a positive effect of using these techniques in food systems (e.g., fruits and vegetables, meat products, and dairy products). However, in spite of the positive effects, some researches showed the negative impacts of using these methods on different food quality aspects such as nutritional, textural, and sensorial properties. We have devoted our attention in this book not only on the advantages of using innovative processing methods but also on the disadvantages and challenges of using these techniques on food quality. This book is designed to assist food scientists as well as those working in the food, nutraceutical, pharmaceutical, and beverage industries. The topics covered in this book are suitable for teaching in courses such as food processing, food chemistry, food biochemistry, sensory science, and new product development. We gratefully appreciate the contribution of all colleagues from all around the world.

Also, we would like to thank Arezou Sobhani from the Department of English and Linguistics, University of Otago, for English proofreading, and the professional assistance provided by the staff of Springer Nature.

St. Paul, MN, USA — Shahin Roohinejad
Compiègne, France — Mohamed Koubaa
Karlsruhe, Germany — Ralf Greiner
St. Paul, MN, USA — Kumar Mallikarjunan

Contents

About the Authors

Shahin Roohinejad obtained his BSc in 2000 in the field of food science and technology from the Islamic Azad University, Iran. He completed his MSc in 2009 in food biotechnology at the Universiti Putra Malaysia (UPM). In July 2011, he received a full doctoral scholarship award from the Department of Food Science at the University of Otago in New Zealand and graduated in December 2014. In June 2015, he received Georg Forster Research Fellowship Award granted by the Alexander von Humboldt Foundation to pursue his postdoctoral research at the Department of Food Technology and Bioprocess Engineering, Max Rubner-Institut (MRI), the German Federal Research Institute of Nutrition and Food. In October 2017, he started to work as the Postdoctoral Research Associate at the Food Science and Nutrition Department, University of Minnesota, USA. Currently, he is a Research Scientist at the Tillamook County Creamery Association in Oregon, USA, a Professional Member of the Institute of Food Technologists (IFT), a Graduate Member of the New Zealand Institute of Food Science and Technology (NZIFST), an Associate Newsletter Editor of the IFT Nonthermal Processing Division (NPD), a Member of IFT Press Advisory Group, and a GHI (Global Harmonization Initiative) Ambassador in the USA. In the last 10 years, he has been working on different food areas such as emerging food processing, emulsion-based systems, nanotechnology, and functional foods. His research activities have resulted in more than 90 original papers in peer-reviewed journals, book chapters, abstracts, and short papers in congress proceedings.

Mohamed Koubaa is an Assistant Professor in the process and chemical engineering at ESCOM Chimie (Ecole Supérieure de Chimie Organique et Minérale) in Compiègne (France). In 2007, he obtained an engineering diploma in biological sciences from the National School of Engineers of Sfax (Tunisia), followed by a master's degree from the University of Technology of Compiègne (France). In July 2008, he received a full doctoral scholarship award from the Ministry of Higher Education, Research, and Innovation (France) and graduated in February 2012. He has performed different postdoctoral research stays as Research Assistant in the Department of Industrial Process Engineering, University of Technology of

Compiègne (France), the Department of Biological Engineering, National School of Engineers of Sfax (Tunisia), and as a Postdoctoral Research Assistant in the prestigious Department of Molecular Genetics, The Ohio State University, Columbus, Ohio (USA). His research focus is on the use of non-conventional processing for the preservation and/or the extraction of bioactive compounds from liquid and solid foods and more recently in microbial growth stimulation using emerging technologies. He has more than 100 publications (book chapters and articles in international journals with peer review).

Ralf Greiner joined the Federal Research Centre for Nutrition, Karlsruhe, Germany, in 1990 as a PhD Student after graduating in Chemistry at the University of Stuttgart. In the early stages of his career as Deputy Head of the Centre for Molecular Biology, he was mainly engaged in research in respect to genetically modified food and enzymes for food processing, with phytases in the center of his interests. In 2007, he held a position as a Visiting Professor for Biochemistry and Molecular Biology, Department of Bioprocess Engineering, Federal University of Paraná, Curitiba, Brazil, working on solid-state fermentation and fungal enzyme production. In 2008, he returned to Karlsruhe where he became the Head of the Department of Food Technology and Bioprocess Engineering of the Max Rubner-Institut (MRI). His research is focused on studying and modeling conventional and new processing technologies, as well as on food nanotechnology, but phytases are still in the focus of his interests. He is a Representative of MRI in several international and national associations on food technology, food control, and food nanotechnology. In 2012, he accepted the position as an Honorary Assistant Professor in the School of Biological Sciences of the University of Hong Kong. His research activities have resulted in approximately 120 original papers in peer-reviewed journals, 37 book chapters, and 260 abstracts or short papers in congress proceedings. In addition, he is the Editor of *Food Control.*

Kumar Mallikarjunan has over 30 years of expertise in food engineering. His research is focused on food process development, process optimization, and nondestructive sensing. He completed his PhD in 1993 from the University of Guelph in Biological Engineering. He worked as a Postdoc at the University of Georgia and as a Faculty in Biological Systems Engineering Department of Virginia Tech. He is currently working with Food Science and Nutrition Department at the University of Minnesota. He has published over 60 refereed research articles and is responsible for research activities over 7 million dollars and has guided 10 PhD and 30 Masters students. In addition, he has published 3 books, more than 13 book chapters, and more than 300 conference presentations in national and international meetings. He also worked closely with industrial projects with companies like Frito-Lay, PepsiCo, Conagra, and Schwan's, to name a few.

Contributors

Min Addy Center for Biorefining, and Department of Bioproducts and Biosystems Engineering, University of Minnesota Twin Cities, St. Paul, MN, USA

Lilia Ahrné Department of Food Science, University of Copenhagen, Copenhagen, Denmark

Erik Anderson Center for Biorefining, and Department of Bioproducts and Biosystems Engineering, University of Minnesota Twin Cities, St. Paul, MN, USA

George Amponsah Annor Department of Food Science and Nutrition, University of Minnesota, St. Paul, MN, USA

Arianna Binello Università di Torino, dipartimento di Scienza e Tecnologia del Farmaco, Torino, Italy

Mladen Brnčić Faculty of Agriculture, University of Zagreb, Zagreb, Croatia

Dongjie Chen Center for Biorefining, and Department of Bioproducts and Biosystems Engineering, University of Minnesota Twin Cities, St. Paul, MN, USA

Yanling Cheng Center for Biorefining, and Department of Bioproducts and Biosystems Engineering, University of Minnesota Twin Cities, St. Paul, MN, USA

Paul Chen Center for Biorefining, and Department of Bioproducts and Biosystems Engineering, University of Minnesota Twin Cities, St. Paul, MN, USA

Giancarlo Cravotto Università di Torino, dipartimento di Scienza e Tecnologia del Farmaco, Torino, Italy

Ivonne Delgadillo QOPNA, Department of Chemistry, University of Aveiro, Aveiro, Portugal

Ricardo V. Duarte QOPNA, Department of Chemistry, University of Aveiro, Aveiro, Portugal

CBQF/Faculty of Biotechnology, Catholic University of Portugal, Porto, Portugal

Liliana G. Fidalgo QOPNA, Department of Chemistry, University of Aveiro, Aveiro, Portugal

Hadi Hashemi Gahruie Department of Food Science and Technology, School of Agriculture, Shiraz University, Shiraz, Iran

Biomolecular Engineering Laboratory, Department of Food Science and Technology, School of Agriculture, Shiraz University, Shiraz, Iran

Ana M. P. Gomes CBQF/Faculty of Biotechnology, Catholic University of Portugal, Porto, Portugal

Ralf Greiner Department of Food Technology and Bioprocess Engineering, Max Rubner-Institut, Federal Research Institute of Nutrition and Food, Karlsruhe, Germany

Sara Hedayati Department of Food Science and Technology, School of Agriculture, Shiraz University, Shiraz, Iran

Rita S. Inácio QOPNA, Department of Chemistry, University of Aveiro, Aveiro, Portugal

CBQF/Faculty of Biotechnology, Catholic University of Portugal, Porto, Portugal

Amaia Lasagabaster AZTI, Department of Food Quality, Safety and Identity, Technological Park Bizkaia, Derio (Bizkaia), Spain

María Lavilla AZTI, Department of Food Quality, Safety and Identity, Technological Park Bizkaia, Derio (Bizkaia), Spain

Guanchen Liu Department of Food Science, University of Copenhagen, Copenhagen, Denmark

Juer Liu Center for Biorefining, and Department of Bioproducts and Biosystems Engineering, University of Minnesota Twin Cities, St. Paul, MN, USA

Iñigo Martínez-de-Marañón AZTI, Department of Food Quality, Safety and Identity, Technological Park Bizkaia, Derio (Bizkaia), Spain

Yiwei Ma Center for Biorefining, and Department of Bioproducts and Biosystems Engineering, University of Minnesota Twin Cities, St. Paul, MN, USA

Silvia A. Moreira QOPNA, Department of Chemistry, University of Aveiro, Aveiro, Portugal

CBQF/Faculty of Biotechnology, Catholic University of Portugal, Porto, Portugal

Daniel Munk Department of Food Science, University of Copenhagen, Copenhagen, Denmark

Mehrdad Niakousari Department of Food Science and Technology, School of Agriculture, Shiraz University, Shiraz, Iran

Jeanette Otte Department of Food Science, University of Copenhagen, Copenhagen, Denmark

Peng Peng Center for Biorefining, and Department of Bioproducts and Biosystems Engineering, University of Minnesota Twin Cities, St. Paul, MN, USA

Zihan Qin Department of Food Science, University of Copenhagen, Copenhagen, Denmark

Rui P. Queirós QOPNA, Department of Chemistry, University of Aveiro, Aveiro, Portugal

Shahin Roohinejad Department of Food Science and Nutrition, University of Minnesota, St. Paul, MN, USA

Roger Ruan Center for Biorefining, and Department of Bioproducts and Biosystems Engineering, University of Minnesota Twin Cities, St. Paul, MN, USA

Nanchang University, Nanchang, China

Mauro D. Santos QOPNA, Department of Chemistry, University of Aveiro, Aveiro, Portugal

Jorge A. Saraiva QOPNA, Department of Chemistry, University of Aveiro, Aveiro, Portugal

Charles Schiappacasse Center for Biorefining, and Department of Bioproducts and Biosystems Engineering, University of Minnesota Twin Cities, St. Paul, MN, USA

Jana Šic Žlabur Faculty of Food Technology and Biotechnology, University of Zagreb, Zagreb, Croatia

Shualing Yang Department of Food Science, University of Copenhagen, Copenhagen, Denmark

Yaning Zhang Center for Biorefining, and Department of Bioproducts and Biosystems Engineering, University of Minnesota Twin Cities, St. Paul, MN, USA

The Harbin Institute of Technology, Haerbin Shi, Heilongjiang Sheng, China

Nan Zhou Center for Biorefining, and Department of Bioproducts and Biosystems Engineering, University of Minnesota Twin Cities, St. Paul, MN, USA

Chapter 1
Impact of Ohmic Processing on Food Quality and Composition

Mehrdad Niakousari, Sara Hedayati, Hadi Hashemi Gahruie, Ralf Greiner, and Shahin Roohinejad

1.1 Introduction

One of the novel promising technologies to pasteurize, sterilize, or cook a wide range of food products (e.g. fruits and vegetables, dairy products, and meat products) is ohmic heating, which usually helps to obtain a high product quality (Fig. 1.1). Ohmic heating is literally an electric resistance heating method, in which an alternating current (50 Hz to 100 kHz) is passed through the food material. In a flow through the unit, several electrode arrangements such as a parallel plate, colinear, parallel rod, and staggered rod electrodes are used depending on the required operation. However, alternative electrode arrangements could also be applied. Similar to other volumetric heating methods, the elimination of heating surfaces in contact with foodstuff in ohmic heating helps to reduce thermal degradation and consequently an enhancement in product quality.

Conventional heating methods provide heating from a hot surface via conductive and convective heat transfer (Damyeh, Niakousari, Saharkhiz, & Golmakani, 2016;

M. Niakousari (✉) · S. Hedayati
Department of Food Science and Technology, School of Agriculture, Shiraz University, Shiraz, Iran
e-mail: niakosar@shirazu.ac.ir

H. H. Gahruie
Department of Food Science and Technology, School of Agriculture, Shiraz University, Shiraz, Iran

Biomolecular Engineering Laboratory, Department of Food Science and Technology, School of Agriculture, Shiraz University, Shiraz, Iran

R. Greiner
Department of Food Technology and Bioprocess Engineering, Max Rubner-Institut, Federal Research Institute of Nutrition and Food, Karlsruhe, Germany

S. Roohinejad
Department of Food Science and Nutrition, University of Minnesota, St. Paul, MN, USA

S. Roohinejad et al. (eds.), *Effect of Emerging Processing Methods on the Food Quality*, https://doi.org/10.1007/978-3-030-18191-8_1

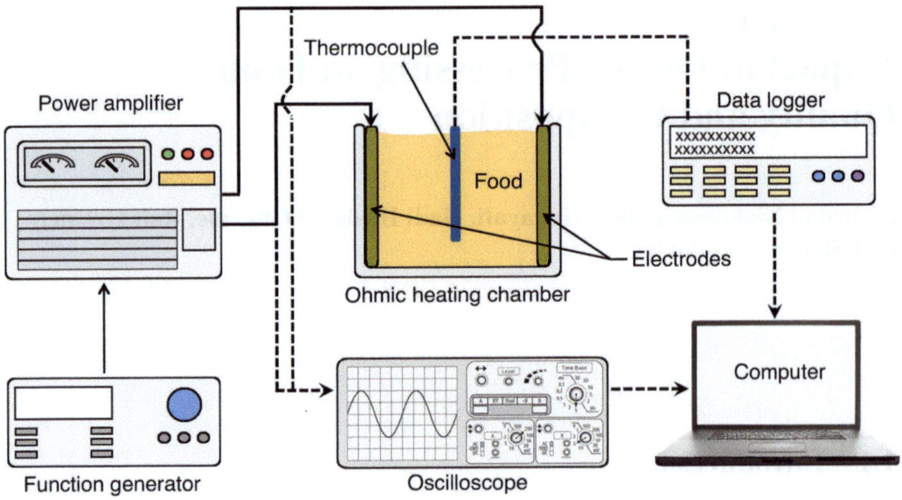

Fig. 1.1 Schematic representation of Ohmic heating equipment for food applications. (Reprinted with permission from Koubaa, Roselló-Soto, Barba-Orellana, and Barba (2016))

Seidi Damyeh & Niakousari, 2016). A hot medium (e.g. steam) in a heat exchanger (e.g. tube, shell, or plate type) or a vessel with a heated jacket can provide heat. These hot surfaces should be at a considerably higher temperature than the food material in order to reach the proper temperature gradient to transfer heat to the food material. As a consequence, thermally degradation or even burning effects could occur, which in turn leads to a reduction in product quality. Furthermore, limited heat transfer, when processing highly viscous fluids and fluids with particulates, causes a very slow process and non-uniform temperature profiles. However, with the aid of mechanical agitation, the required heat transfer and uniformity would be obtained, which may cause physical damages to the product (Niakousari, Hashemi Gahruie, Razmjooei, Roohinejad, & Greiner, 2018).

Application of ohmic heating overcomes the aforementioned downsides of conventional heating via direct heat generation inside the food material (Seidi Damyeh & Niakousari, 2017). Therefore, enhanced product quality (i.e. color, flavor, nutrient retention) is obtained owing to the absence of thermal degradation. Low flow rates or high viscosity fluids would not contribute to the performance decline in ohmic heating, which is because of not being dependent on turbulence effects. Ohmic heating also obviates the need for mechanical shear in particulate food material, which may cause damage to the product that is obviously due to its uniform heating.

In order to be successfully processed via ohmic heating, the food material should be pumpable with appropriate electrical conductivity (0.01–10 S/m). High process efficiency is obtained by ohmic heaters, in which over 95% of electrical energy is transformed to heat within the food material (in 50 Hz systems) (Damyeh, Niakousari, Golmakani, & Saharkhiz, 2016). This technique offers the ability to efficiently control the process and to have a rapid start-up due to its electrical nature. Nevertheless, the cost of electrical energy can be considered as a limitation to ohmic heating.

Milk pasteurization was the very first application of ohmic heating developed in the early twentieth century (Anderson & Finkelstein, 1919). This process was formerly known as "electropure process" and was stopped using due to lack of appropriate electrode material. By development of sterilization process of food materials having large particulates at the United Kingdom Electricity Council Research Center, more research into ohmic heating was recommended in the 1980s (Simpson, 1983; Simpson & Stirling, 1995; Stirling & Coombes, 1990). An enhancement in quality regarding particulate identity of food's color, and vitamin retention and longer shelf life for pasteurized food products by the industrial application of ohmic heating (compared to conventional heating) has been globally observed since the 1980s (Anderson, 2008; Castro, Teixeira, Salengke, Sastry, & Vicente, 2004; Leizerson & Shimoni, 2005a). A high-quality product with minimal structural, organoleptic and nutritional changes was produced by Rahman (2007) in a short operating time using ohmic heating. In this chapter, the influence of ohmic heating on the nutritional, textural and sensorial attributes of different foodstuffs such as fruits and vegetables, dairy, and meat products were highlighted and the advantageous and disadvantageous of using this technology on food quality were discussed.

1.2 Nutritional Properties

The time required to increase the cold point temperature in conventional heating may over process other parts of the food, which decreases the nutritional and sensory properties of food products. Emerging thermal treatments such as ohmic heating can be used to overcome this drawback of conventional heating operations by homogenous heating distribution throughout the products. The influence of ohmic processing on nutritional properties of different food products is discussed below.

1.2.1 Fruits and Vegetables

The nutritional interest of fruits and vegetables is due to the presence of high levels of fibers, vitamins, minerals, phenolic compounds and bioactive peptides (Septembre-Malaterre, Remize, & Poucheret, 2018). The degree of these components in fruits and vegetables depends on their variety, maturity, agronomical practices, postharvest operations and processing conditions such as heating operations. Vitamin C (ascorbic acid) is essential for the synthesis of collagen and its deficiency causes scurvy. Moreover, ascorbic acid is a natural antioxidant applied in the food industry to impede browning, discoloring and to improve the shelf life of food products. However, it is considered a heat-sensitive vitamin. The degradation of ascorbic acid in foods is influenced by factors such as oxygen concentration, light, temperature, pH, a_w (water activity) and the presence of metallic ions (Damodaran, Parkin, & Fennema, 2008). Several studies have evaluated the thermal degradation of vitamin C under different heating conditions including emerging technologies such as ohmic heating. The results of some of these studies are listed in Table 1.1.

Based on these results the influence of ohmic heating on vitamin C degradation depends on process conditions such as the voltage gradient and duration of ohmic heating. However, contradictory results have been reported for different products. These differences may be due to the dissimilarities in electrochemical reactions in these products. Nevertheless, in frequencies above 100 Hz, the occurrence of these reactions decreased, and degradation kinetics of ascorbic acid was not influenced by high electric field frequency. This shows that hydrogen donation tendency of the ascorbic acid molecule in redox reactions is not influenced by the rapid changes in the electric field (Mercali et al., 2014). In addition to the voltage gradient, the type of electrode has been proven to have significant effects on vitamin C degradation. Athmaselvi et al. (2017) determined the effect of titanium and stainless steel electrodes on ascorbic acid degradation in the tropical fruits (papaya, sapota, and guava) under ohmic heating and found that samples heated with titanium electrode exhibited better ascorbic acid retention. Louarme and Billaud (2012) studied the impact of conventional heating and ohmic heating on the degradation of vitamin C in apple and peach desserts. The results showed that in contrast to conventional heating, the ohmic heating had very little effect on oxidative degradation of ascorbic acid. In another study, Jaeschke, Marczak, and Mercali (2016) evaluated the influence of the

Table 1.1 Influence of ohmic heating on ascorbic acid degradation of fruits

Product type	Process condition	Result	Reference
Aonla pulp	Samples heated for 1 min at 90 °C at 11, 13, 15 or 17 V/cm	Low voltage gradients brought more ascorbic acid degradation	Singh et al. (2013)
Guava juice	Samples heated for 1, 3 and 5 min at 95 °C at 13.33, 16.66, 20 and 23.33 V/cm voltage gradient at 50 Hz frequency	High voltage gradients induced more ascorbic acid degradation	Chakraborty and Athmaselvi (2014)
Sweet lime juice	Samples heated with a temperature range of 27.4 °C to 85.4 °C at 30, 40 and 50 60 V/cm voltage gradient at 50 Hz frequency	High voltage gradients induced greater ascorbic acid degradation	Parmar, Tripathi, Tiwari, and Singh (2016)
Papaya, sapota, and guava	–	High voltage gradients induced greater ascorbic acid degradation	Athmaselvi, Kumar, and Poojitha (2017)
Acerola pulp	Samples heated for 3 min at 85 °C at a heating voltage of 120–200 V	High voltage gradients induced greater ascorbic acid degradation	Mercali, Jaeschke, Tessaro, and Marczak (2012)
	Samples were heated for 0, 20, 40, 60, 80, 100 and 120 min at 10, 10^2, 10^3, 10^4 and 10^5 Hz	Low voltage gradients induced greater ascorbic acid degradation	Mercali, Schwartz, Marczak, Tessaro, and Sastry (2014)

electric field on the degradation of carotenoid and ascorbic acid in acerola pulp at different temperatures (80, 85, 90 and 95 °C) during 60 min of heat treatment (ohmic or conventional heating). Different temperatures exhibited the same effects on degradation rates and the degradation process was similar in both heating methods, indicating that the electric field did not affect the mechanisms and rates of ascorbic acid and carotenoid degradation.

When a sample is exposed to an electric field, its molecules align with the oscillating electric field in a process known as polarization (Mercali et al., 2014). In acerola fruit, the polarization process did not influence the rate of ascorbic acid degradation in ohmic heating. This technology was performed under atmospheres with low or high oxygen contents at 90 °C in order to explore the impact of oxygen on the ascorbic acid and carotenoid degradation. They found that the degradation of carotenoid was greater under a rich oxygen atmosphere (Mercali et al., 2012). Similarly, Mercali et al. (2014) studied the effect of electric field frequencies (between 10 and 105 Hz) on the degradation of ascorbic acid in acerola pulp during ohmic heating and compared the results with conventional heating. Low frequency caused greater ascorbic acid degradation possibly because of the incidence of electrochemical reactions. However, these reactions were minimized at frequencies above 100 Hz and degradation rates of ascorbic acid were the same as for conventional heating.

Carotenoids are great antioxidants and some of them (e.g. carotene) have been proven to offer provitamin A activity. Carotenoid degradation occurring by isomerization or oxidation is the main reaction to its degradation, which depends on carotenoid structure (Rodriguez-Amaya, 2001). Achir et al. (2016) explored the influence of ohmic and conventional pasteurization on carotenoid protection in grapefruit and blood orange. The results revealed that xanthophyll losses reached 70% for epoxyxanthophylls and 40% for hydroxyxanthophylls in conventional heating, while losses were below 30% and 20% in ohmic pasteurization, respectively. Lycopene and β-carotene (carotene species) were stable in both heating treatments. Thus, ohmic pasteurization was suggested as a promising option for the preservation of xanthophylls. In another study, the vitamin degradation kinetics and changes in the carotenoid contents of orange juice samples heated by four different methods were investigated and better vitamin retention at all temperatures was reported using ohmic heating (Vikram, Ramesh, & Prapulla, 2005).

Mercali, Gurak, Schmitz, and Marczak (2015) studied the non-thermal impacts of electricity on the degradation of anthocyanin, in ohmic heating of jaboticaba juice and found that electric field did not influence the degradation rate of anthocyanins. Sarkis, Jaeschke, Tessaro, and Marczak (2013) compared the influence of conventional and ohmic heating of blueberry pulp on its anthocyanin degradation. They found that the degree of anthocyanin degradation under lower voltages was lower or comparable to those obtained during conventional heating. However, ohmic heating under high electric fields exhibited higher anthocyanin degradation. In the research conducted by Guida et al. (2013), the nutritional properties of conventional and ohmic heated artichokes were compared. The conventional cooking (boiling) or blanching of artichokes resulted in the loss of nutritional properties and reduction of antioxidant compounds owing to thermal degradation, oxidation, and leaching (Lutz, Henríquez, & Escobar, 2011). They also found that the rate of peroxidase and polyphenol oxidase inactivation in conventional heating was slower and phenolic compounds were degraded during heating. However, after ohmic heating, the total phenolic content was increased which might be due to the release of bound phenolics and also the breakdown of artichoke cells that increased the accessibility of the antioxidant compounds. With the increase of electric field strength (from 25 to 30 and 40 V/cm), the retention of total phenolic content was higher due to the faster polyphenol oxidase inactivation. Castro, Macedo, Teixeira, and Vicente (2004) determined the inactivation kinetics of pectinase, lipoxygenase, alkaline phosphatase, polyphenoloxidase, and β-galactosidase under ohmic and conventional heating. The thermal history in both heating techniques was made equal to investigate the enzyme inactivation caused by the electric field. The results showed that all of the enzymes in both heating operations followed first-order inactivation kinetics. They also found that the electric field did not influence the inactivation rate of β-galactosidase, pectinase, and alkaline phosphatase. However, the inactivation of polyphenoloxidase and lipoxygenase were accelerated in ohmic heating. The electric field removed the metallic prosthetic groups in lipoxygenase and polyphenoloxidase, changed the molecular spacing and increased the interchain reactions and facilitates the inactivation of these enzymes. Therefore, the ohmic-treated samples were demonstrated to present higher levels of antioxidants and phenolic compounds compared with conventional-heated ones.

Several studies have investigated the changes occurred in proteins and amino acids of fruits and vegetables during ohmic heating. For instance, Mesías, Wagner, George, and Morales (2016) assessed the impact of ohmic and conventional retort sterilization on protein quality and amino acid content of sterilized vegetable baby foods. The outcomes showed that heating did not influence the total content in proteins. Nevertheless, after retort sterilization, the degree of total essential and non-essential amino acids was decreased significantly. On the contrary, ohmic heating did not show a significant impact on the extent of amino acid and can be used instead of conventional sterilization in vegetable baby foods in order to preserve the nutritional quality of proteins. Moreno et al. (2017) enriched apples with L-arginine under ohmic heating, vacuum impregnation/ohmic heating or conventional heating. The greater L-arginine addition was obtained in vacuum impregnation/ohmic heating followed by ohmic treatment and in conventional treatment, less addition of L-arginine was observed. This behavior was suggested to be associated with the acceleration of mass transfer in apple particles due to osmotic diffusion and hydrodynamic mechanisms. In another study, the protein content of canned artichokes blanched with conventional and ohmic heating was measured by Guida et al. (2013). They reported that protein content was decreased with storage time in ohmic and conventional cooked samples, but a higher rate protein loss was observed in conventionally cooked samples. After three-month storage, the proteins loss in conventional and ohmic blanched samples was 74.1% and 41.3%, respectively.

Apart from the preservation of nutrients, ohmic heating has shown to prevent the formation of carcinogenic compounds such as furan. Hradecky, Kludska,

Belkova, Wagner, and Hajslova (2017) investigated the furan formation in ohmic and conventional retort sterilization of baby foods (vegetable and vegetable/meat). Furan was significantly decreased (70–90%) in ohmic-heated samples due to decreased degradation of furan precursor in ohmic heating. The fatty acids oxidation and Maillard reaction products were greater in samples sterilized by conventional heating compared to ohmic-heated samples.

1.2.2 Meat Products

Meat is a substantial part of our diet and provides protein, minerals, vitamins, fatty acids, and energy (Hashemi Gahruie et al., 2017; Toldrá, 2017). The proteins of meat have high biological value due to the similarity of amino acid (types and proportions) to those in human muscles. Moreover, meat contains all of the eight essential amino acids (lysine, leucine, isoleucine, methionine, phenylalanine, valine, threonine, and tryptophan), which are not synthesized in the human body (Purchas, Wilkinson, Carruthers, & Jackson, 2014). Oxidation of amino acids results in the development of carbonyl groups, which react with free amines of proteins to form amide bonds. These reactions negatively affect the nutritional quality and digestibility of meat proteins (Gatellier, Kondjoyan, Portanguen, & Santé-Lhoutellier, 2010).

Dai et al. (2014) measured the carbonyl contents of ohmic and water bath cooked pork meat under refrigerated storage. No significant differences were observed between ohmic and water bath cooked meat and all of the samples contained the same levels of carbonyls. Protein oxidation was also attributed to the reduction of free thiol groups, which were changed into disulfide bonds. The formation of disulfides promoted protein aggregation and negatively affected the nutritional quality of meat (Gatellier et al., 2010; Lund, Lametsch, Hviid, Jensen, & Skibsted, 2007; Santé-Lhoutellier, Astruc, Marinova, Greve, & Gatellier, 2008).

Cooking has remarkable effects on nutritive values of meat due to the occurrence of physicochemical reactions. One of the most important factors affected by heating is the oxidation of lipids, which negatively affects the quality and acceptability of meat products (Fuentes, Ventanas, Morcuende, Estévez, & Ventanas, 2010). Lipid oxidation promotes the formation of free radicals that are responsible for myoglobin oxidation (Kim, Huff-Lonergan, Sebranek, & Lonergan, 2010). Moreover, oxidation of fatty acids leads to the formation of a variety of low aroma threshold products such as aldehydes and ketones. These volatile compounds lead to off-flavor in cooked meat (Toldrá, 2017). Dai et al. (2014) determined the thiobarbituric acid reactive substances (TBARS)-values of ohmic and water bath cooked meat samples. The ohmic heated samples showed lower levels of lipid oxidation and a slower rate of increase in oxidation during storage. Non-heme iron, one of the main catalysts of lipid oxidation, was also lower in ohmic-cooked samples. Zell, Lyng, Cronin, and Morgan (2010a) reported similar findings for turkey samples after cooking by ohmic and conventional heating processes. Their results showed that ohmic-treated samples had lower lipid oxidation and sulfur-flavor-compound compared to conventional-heated samples. This might be due to the prolonged exposure of the outer layers of turkey meat to high temperature, denaturation of myofibrillar proteins and thermal degradation of membrane phospholipids under conventional cooking. In another study, δ-tocopherol was incorporated into beef patties cooked with impingement and ohmic heating. To control lipid oxidation, samples were assessed using an ethanolic carrier (Wills, Dewitt, Sigfusson, & Bellmer, 2006). The results revealed that ohmic-cooked samples contained significantly higher TBARS than impingement-cooked samples. Patyukov and Pacinovski (2015) applied conventional and ohmic heating on polyunsaturated fatty acids (PUFA) fortified cooked sausages. The results showed that conventional cooking led to an obvious deterioration of fat in sausages. Triglycerides were hydrolyzed and PUFA were oxidized during conventional heating. However, short heating time using ohmic technology decreased the deterioration of fatty acids and ohmic-treated samples showed significantly lower oxidation. The settings of the ohmic heater during heating had significant influences on the oxidation of lipids. In a study performed by Kim, Hong, Park, Spiess, and Min (2006), frozen pork patties were ohmically thawed under different power intensities and found that the TBARS increased proportionally to the power intensity, and induced lipid oxidation in pork patties.

Thawing is a critical stage to guarantee the microbial and nutritional value of food products. Several studies have confirmed that this operation must be carried out rapidly to prevent microbial growth and nutritional losses caused by leaching of soluble proteins (Roberts, Balaban, Zimmerman, & Luzuriaga, 1998). Duygu and Ümit (2015) compared the quality of meat samples thawed in a refrigerator, at room temperature or ohmic heating. The fastest thawing was obtained in the samples treated with ohmic heating. An increase in pH value and a decrease in a_w were observed with all thawing methods. However, the ohmic-thawed samples had the highest a_w value. Water activity depends on water loss and ohmic heating resulted in the least weight loss. Since the loss of water-soluble vitamins and proteins were in coincidence with water loss, ohmic thawing could preserve the nutritional quality of thawed foods. Thus, ohmic heating was suggested as an efficient technique in thawing. The influence of ohmic heating on heavy metals has also shown promising results. Bastías et al. (2015) assessed the influence of ohmic heating and blanching at different temperatures (50, 70 and 90 °C) on lead (Pb) and cadmium (cd) content in Chilean blue mussel (*Mytilus chilensis*). The results showed that Pb and Cd content in fresh mussels were reduced by ohmic heating and the greatest reduction was observed in ohmic-treated samples at 90 °C.

1.2.3 Dairy Products

Dairy products are great sources of beneficial nutrients such as minerals and vitamins, which improve bone health, prevent dental cavities and osteoporosis. They also reduce the blood pressure and risk of colon cancer (Wells, 2001). These nutritional benefits make milk an important food item in our daily diet. The composition of fatty acid in milk, predominantly polyunsaturated fatty acids (PUFA) have also significant effects on nutritional properties of dairy products (Hurtaud & Peyraud, 2007). Heating conditions may have substantial influences on the extent of these components. Similarly, the findings of Pereira, Martins, and Vicente (2008) revealed that the profile of short-chain and medium-chain free fatty acid was not affected by heating operation and no significant variations were observed between raw, conventional and ohmic pasteurized milk. Pereira et al. (2016) determined the effects of ohmic heating on unfolding, denaturation and aggregation kinetics of WPI proteins. Their results showed that ohmic heating increased the formation of linear structured protein aggregates and their size was impacted by the electric field intensity. They also found that these aggregates could be used for the encapsulation of bioactive compounds. In another study, Roux et al. (2016) used steam injection and ohmic heating for sterilization of liquid infant formula and compared their effects on soluble proteins and vitamin C. The degree of soluble protein denaturation was similar for both heating methods, while ohmic heating better-preserved vitamin C. Irudayaraj, McMahon, and Reznik (2000) stated that the extent of protein denaturation in the ohmic-sterilized milk was significantly lower than commercial UHT milk. Castro, Macedo, et al. (2004) examined the influence of conventional and ohmic heating on milk proteins using polyacrylamide gel electrophoresis (PAGE). The results showed that raw, conventional and ohmic treated samples presented the same electrophoresis pattern. Roux, Courel, Ait-Ameur, Birlouez-Aragon, and Pain (2009) investigated the occurrence of Maillard reaction in a model infant formula under ultra-high temperature (UHT) by ohmic heating. Five heating temperatures (100–140 °C) were selected and samples were taken during the operation (heating and holding) and examined for early Maillard reaction products (MRPs), such as furosine, advanced products such as advanced glycated end products (AGEs) and carboxymethyllysine (CML) and final products such as melanoidins. The results showed that Maillard reactions in model infant formula were highly dependent on time-temperature treatment and the amount was increased with ohmic heating temperature and time.

1.3 Textural Properties

The texture is one of the most significant quality characteristics of foods. The textural parameters of food products such as fruits, vegetables, meat and meat products, cereal-based foods, and cheeses are determinant in palatability of these food materials. They have also remarkable influences on their price. Hence, reaching the desired textural quality of food products has major financial importance. Usually, the food processing operations are designed to change the textural properties of food products and dominantly these changes are directed to weaken the food structure and make it easier to masticate. However, sometimes the textural changes are inadvertent and usually undesirable (Bourne, 2002). Ohmic heating technology has been extensively investigated and used in a variety of food processing operations and the results have revealed that it can modify the textural properties of food products.

1.3.1 Fruits and Vegetables

The structural integrity and textural quality of fruits and vegetables are mostly related to the middle lamella, cell wall and the turgor within cells (Waldron, Parker, & Smith, 2003). Among these parameters, plant cell walls play an important role in the textural properties of plant-based foodstuff (Van Buren, 1979). The plant cell wall consists of three major polysaccharides namely cellulose, hemicellulose, and pectin. There are also varying levels of structural proteins and phenolic compounds in the cell wall. The cellulose provides the microfibrillar component of the cell wall, while hemicelluloses are bound to cellulose and pectin by hydrogen linkages. Postharvest operations and storage have dramatic influences on structural integrity and biochemical composition of plant cells. During most of the food processing operations, the texture firmness and structural integrity of cells are lost and lead to loss of turgor and crispiness. Generally, processing or storage does not notably influence the firmness caused by the cellulose and hemicellulose. However, pectin could be influenced by the enzyme-catalyzed reactions during thermal treatment, which causes remarkable changes in fruit and vegetable products.

Ohmic heating leads to obvious damage to the cell membranes and enhances the mass transfer processes. Moreover, membrane rupture significantly increases the electrical conductivity of plant tissues and influences the heating process. Shynkaryk, Ji, Alvarez, and Sastry (2010) investigated the ohmic heating of peaches at an electric field strength of 60 V/cm and frequencies between 50 Hz and 1 MHz. The textural measurements of samples heated at lower frequencies showed a significant loss of textural strength. They stated that the application of a low-frequency electric field causes the cell membrane electroporation, increases the electrical conductivity of the product, and increases the texture-softening rate. Generally, the texture relaxation data showed even higher stages of tissue damage at low frequency. Thus, high-frequency ohmic heating would minimize the texture degradation of peach tissue. They reported that combination of operation parameters (200 kHz at 60 V/cm to 65 °C for 8 s) minimized the textural damage of peaches during the thermal preservation. The study carried out by Olivera, Salvadori, and Marra (2013) focused on the influences of ohmic heating on textural properties of fresh potatoes, carrots and apples under 1100, 2200, and 3300 V/m electrical field gradient. The authors compared the results with the raw untreated samples. They cut the samples into cylinders (h = 9.0 mm, d = 30 mm) and ohmically heated them for 60, 120, 180 or 240 s. The compression test of the samples treated by ohmic heating differed from raw samples for all cooking conditions and firmness was reduced with the heating time. The electric field intensity had also a great influence on the textural properties. In potato and carrot, significant firmness disintegration was observed only for the electric field strength of 2200 V/m and higher. However, apple samples were more sensitive and exhibited an obvious firmness-decreasing trend with the electric field strength. The differences in texture softening of these products can be described by the differences in their tissue structure, cell size and the amount of air cavities.

The utilization of ohmic heating for blanching of artichokes was investigated by Guida et al. (2013) and compared with conventional heating. The authors concluded that the texture of the artichokes blanched by the conventional method was not homogenous. The core areas showed lower texture softening but the outer part was over-processed and excessively softened. This could be attributed to the non-uniform heating of artichokes due to low heat transfer by conduction in conventional blanching. They also found that the samples treated by ohmic heating offered lower hardness compared to those processed by conventional heating due to higher heating rate in ohmic treatment compared to conventional heating. Kamali and Farahnaky (2015) compared the textural properties of ohmic, microwave and conventionally cooked radish, cabbage, potato and turnip by texture profile analysis. The ohmic heating was performed at 220 and 380 volts. Ohmic heating at 380 V had greater texture softening effects for all of the studied vegetables in the order of radish > turnip > potato > cabbage. In contrast to microwave and conventional cooking, the cohesiveness of samples was increased in ohmic heating with cooking time. Cooking regardless of heating method increased the springiness. However, ohmic heating had greater effects on this parameter, which could be due to limited structural damage Chiavaro, Barbanti, Vittadini, and Massini (2006). Ohmic-cooked turnips presented a more compact structure with smaller pores compared to other heating methods. However, conventional and microwave cooking caused structural damage and wall thinning. In another study by Farahnaky, Azizi, and Gavahian (2012), the effects of ohmic, conventional and microwave heating on the textural parameters of red beet, carrot, and golden carrot were examined by texture profile analysis (TPA) at different heating times. They found that ohmic heating not only caused more texture softening rates but also the final hardness, gradient and compression energy of ohmic-heated samples were significantly lower than those of other treatments. Jittanit et al. (2017) found the same results for rice samples. They compared the ohmic cooking of rice with the conventional heating in an electric rice cooker. They used four types of rice samples including two varieties of white rice (KDML105 and Sao Hai), brown rice, and germinated brown rice (KDML105). They found that it is possible to ohmically cook all types of rice by using salt solution (0.1 M) in the mixtures to increase the electrical conductivity. The electrical energy consumption in ohmic cooking was approximately 73–90% of energy consumed in the electrical rice cooker. The texture of the samples cooked by ohmic heating was meaningfully softer than the samples cooked by electric rice cooker and the extent of difference was dependent on the rice type.

1.3.2 Meat Products

The texture is one of the most important quality characteristics of meat and meat products and depends on some factors such as the properties of myofibrils and connective tissues, the pH, and the fat content and its distribution. The textural properties of meat can be modified significantly by changing the conditions of refrigeration, storage or cooking (Mathoniere, Mioche, Dransfield, & Culioli, 2000). Most heating operations applied in meat processing are based on convective, conductive, and/or radiative mechanisms from a heated medium (e.g. water, air, oil, etc.) to the meat product. However, it may take a long time to conduct enough heat into the core of meat products and it causes over-processing in most of the parts of the product, which negatively affects the quality (Wills et al., 2006). Therefore, during the past few years, alternative heating techniques such as ohmic heating have gained importance in meat processing and several research projects have been performed to evaluate the suitability of ohmic heating for meat and meat products processing. Some of these studies are summarized in Table 1.2. Most of these findings revealed that the ohmic-cooked samples had a firmer texture than the conventional-cooked ones.

The generation of high temperature in ohmic cooking results in more collagen shrinking and toughening of meat (Dai et al., 2014). Greater water release and more extensive protein denaturation in ohmic heating have been reported to increase the firmness of meat products (Zell et al., 2010a). Moreover, ohmic treated samples had more uniform structure compared to the conventional-cooked samples. The main reason for this behavior is that in ohmic heating, the heat is generated throughout the samples homogeneously and a uniform protein network is formed within the product, which results in a firm and uniform microstructure. On the contrary, during conventional cooking of meatballs in hot water, the heat was not transferred to all parts of the sample simultaneously; hence, a temperature gradient was formed within the sample and led to a heterogeneous structure. Engchuan et al. (2014) reported that conventionally cooked pork meatballs had a heterogeneous structure with cracks inside its matrix. Ohmic heating has also been applied for seafood processing and has increased their firmness due to the rapid inactivation of endogenous proteinases, which lead to further retention of undamaged myofibrillar proteins and improves the gel network structure. Also, the uniform unfolding of myofibrillar proteins in ohmic heating improved the textural properties of ohmic-heated seafood (Chai & Park, 2007; Tumpanuvatr & Jittanit, 2012). However, samples treated with lower voltage gradient showed higher breaking force. Apparently, proteinase inactivation is not the only factor affecting the gel structure and gel network also depend on the aggregation of proteins. When high voltages are applied, the heating time is very short, and the gel network will not develop properly, as a result, the textural parameters would decrease (Foegeding, Allen, & Dayton, 1986). In another study, Özkan et al. (2004) studied the cooking of hamburgers with a combination of ohmic and conventional heating as well as conventional heating and observed no significant differences in the quality and the mechanical properties of the obtained samples.

Table 1.2 Influence of ohmic heating on textural properties of meat products

Product type	Process condition	Textural test type	Result	Reference
Frankfurters	In ohmic heating, the voltage gradient was 3 and 5 V/cm at 50 Hz supply. The conventional steam cooking was performed at 80 °C for 2 min	Texture profile analysis (TPA)	No significant difference occurred between samples cooked by steam or ohmic methods in hardness, energy, cohesion, gumminess, and chewiness	Shirsat, Brunton, Lyng, McKenna, and Scannell (2004)
Hamburger patties	In ohmic heating, voltage was remained constant at 50 V and current started from zero and reached 13 A. While in conventional cooking samples were placed between the heated plates of the grill at 180 °C	Compression test	The mechanical properties of the samples cooked by ohmic or conventional heating are very similar	Özkan, Ho, and Farid (2004)
Ground beef	Ground beef samples with different fat contents (2%, 9%, and 15%) were ohmically (20, 30 and 40 V/cm) and conventional heating was performed on the grill (90 °C) for 18 constant periods (90 s) until reaching a center temperature of 70 °C	Warner–Bratzler shear force	Ohmically cooked ground beef cylinders were firmer than conventionally cooked samples and the voltage gradient did not affect the quality of cooked meat. The ohmic cooking time was increased as the fat content increased due to the poor electrical conductivity of fat	Bozkurt and Icier (2010)
Beef cuts	Samples were thawed from 10 to 18 °C by applying different voltage gradients (10, 20 and 30 V/cm) during ohmic treatment whereas conventional thawing was applied at constant temperature (25 °C, 95% RH)	Texture profile analysis	Ohmic thawing resulted to obtain harder thawed beef cuts than conventionally thawed ones	Icier, Izzetoglu, Bozkurt, and Ober (2010)
Turkey meat	Intact turkey meat was cooked using low-temperature long time (LTLT) and high-temperature short time (HTST) protocols in a combined ohmic/convection system and compared to conventional steam cooking	Texture profile analysis (TPA)	No significant differences between ohmic LTLT and mean values of conventional treatments were found, however, the ohmically cooked HTST displayed a significantly firmer texture	Zell et al. (2010a)
Whole beef muscle	Samples were cooked in a combined ohmic/convection heating system to low (72 °C, LTLT) and high (95 °C, HTST) target end-point temperatures. A control was also cooked to an end-point temperature of 72 °C at the coldest point	Warner-Bratzler shear force	Warner-Bratzler shear force values showed no significant differences between treatments indicating a similar level of tenderness amongst the products	Zell, Lyng, Cronin, and Morgan (2010b)

(continued)

Table 1.2 (continued)

Product type	Process condition	Textural test type	Result	Reference
Surimi	For ohmic heating, the samples were heated to 90 °C at a frequency of 10 kHz at voltage levels of 100 and 250 V, corresponding to a voltage gradient of 6.7 and 16.7 V/cm respectively. For water bath heating. Samples were heated at 90 °C for 30 min	Penetration test	Ohmic heating with a voltage gradient of 6.7 resulted in firmer gels compared to 16.7 V/cm and the breaking force and deformation of ohmic heated samples were higher than those heated in a water bath	Tadpitchayangkoon, Park, and Yongsawatdigul (2012)
Pork meatball	Heating until the center temperature reached 74 °C in a boiling water bath at a rate of 4.9 °C/min and 4.9 °C/min or 24.5 °C/min in ohmic heating at 72 V	Yield strength	The texture of ohmically-heated meatball (at the heating rate of 4.9 °C/min) was stronger than that of ohmically-heated (at a heating rate of 24.5 °C/min) and conventionally-heated meatballs and the conventionally-cooked meatballs had the softer texture	Engchuan, Jittanit, and Garnjanagoonchorn (2014)
Pork meat	The pork meat samples were ohmically-cooked to a minimum end-point temperature of 95 °C at the cold spot by applying a 10 V cm^{-1} voltage gradient. While the conventional cooking time was approximately 83 min at 95 °C	TPA analysis	ohmically-cooked samples had significantly higher hardness, chewiness, and gumminess but same values of springiness	Dai et al. (2014)
Shrimp	Shrimps were cooked until the cold-spot reached 72 °C either by steaming in a conventional steamer or in an ohmic heating cell of 50 Hz at 120 V and 15A	Warner-Bratzler shear force (WBSF) and Kramer	The application of ohmic heating to shrimps did not affect their texture when compared to conventional cooking methods	Lascorz, Torella, Lyng, and Arroyo (2016)
	Maximally supply a 230-voltage using alternating current (60 Hz, sinusoidal)	Texture profile analysis (TPA)	No significant changes were seen for the textural parameters of different samples and shrimps achieved a comparable quality compared to conventional heating processes reported in the literature	Pedersen, Feyissa, Brøkner Kavli, and Frosch (2016)

Ohmic heating can be a promising option for thawing and has been proven to bring about fewer textural changes compared to conventional thawing. This behavior can be attributed to the higher denaturation rate in the myofibrillar proteins using the conventional method, which causes softer thawed products (Icier et al., 2010). The histology of beef cuts proposed by Icier et al. (2010) also confirmed these results and more deformation and loss of collagen fibrils were detected in beef cuts thawed by the conventional method. These changes occurred due to the contraction of myofibrils or denaturation of collagen fibril network.

1.3.3 Dairy Products

The application of ohmic heating is expected to have remarkable effects on the textural properties of dairy products. However, only a few papers have explored the influence of this technique on the textural properties of dairy products. The findings of Pereira et al. (2016) showed that the application of ohmic heating on whey protein isolate (WPI) resulted in the formation of linear structured proteins that had the potential to form gels and to be used as a thickening or gelling agent. In another research, Icier (2009) used ohmic and water-bath techniques for the heating of reconstituted whey solutions. It was found that the consistency coefficients of whey solutions decreased by increasing the thermal treatment time in both heating operations. They also reported that fast heating caused a delay in some processes such as the degradation and the gelation. It was concluded that the electrochemical reactions occurring during ohmic heating might be responsible for these textural changes.

1.4 Sensorial Properties

Heating operations such as ohmic heating may induce significant differences in the sensory attributes and acceptability of different food products. For instance, Christian and Leadley (2006) studied the impact of ohmic treatment on the quality of different products such as shelf-stable milk, soups, puddings, fruit juices, fruit concentrates, and liquid egg products. Rapid and uniform heating was observed with a decreased influence on the organoleptic properties of products. The effects of ohmic treatment on the sensorial properties of different food products are discussed in the following sections.

1.4.1 Fruits and Vegetables

Color, flavor, and texture are important quality characteristics influencing the consumer acceptance and sensory perception of fruits and vegetables (Oey, Lille, Van Loey, & Hendrickx, 2008). These parameters may be influenced by heat treatments such as ohmic heating. Tumpanuvatr and Jittanit (2012) compared the sensory attributes of conventionally- heated and ohmic-heated pineapple and orange juices. The results demonstrated that both heating systems deteriorated the sensorial attributes of fruit juices at the analogous level and the quality of processed juices was not dependent on the heating systems. In other words, the existence of an electric field in the ohmic system did not bring about an additional impact on the juice quality. In a research conducted by Dima, Istrati, Garnai, Serea, and Vizireanu (2015), a sensory evaluation by a group of seven panelists evaluated the appearance, color, taste, odor, and mouthfeel of fresh, conventional pasteurized and ohmic pasteurized vegetable juices. Ohmic pasteurized juices obtained a higher score than conventionally pasteurized juices and concluded that ohmic pasteurization has not a negative effect on the flavor of vegetable juices.

Color is one of the most important sensory attributes and has been extensively applied for quality evaluation of food products. Kim, Ryang, Lee, Kim, and Rhee (2017) exposed the apple juice samples to ohmic heating (frequency = 25 kHz, electric field strength = 26.7 V/cm) at 85–100 °C for 30–90 s and compared the results with the conventional heating. They found that the color values (L∗, a∗, and b∗) of apple juice remained almost unchanged in all treatments, but the lightness and yellowness of ohmic-treated samples at 100 °C for 60 s were decreased. Kim et al. (2017) measured color parameters of L∗, a∗, and b∗ and lycopene content to assess the quality of tomato juice. The results showed no significant changes between the color parameters and lycopene content of the treated and untreated samples. Aamir and Jittanit (2017) compared the influence of ohmic heating and conventional heating on Gac aril oil extraction. They reported that the amount of lycopene and β-carotene of Gac aril oil were increased by the extraction with the ohmic heating method. The color parameters of the extracted oil and residue obtained by conventional and ohmic heating were remarkably different from each other. The SEM images showed that ohmic heating caused more rupture on the cell wall of the Gac aril powder compared to conventional heating and the ohmically extracted oil was redder than that of the conventionally extracted one. This was reported to be due to the ability of the ohmic system to extract higher levels of red pigments (lycopene and β-carotene) from the Gac aril to the hexane solvent.

In a study by Jaeschke et al. (2016), the effect of ohmic heating on acerola pulp was studied. Fresh acerola pulp had an orange/red color, but the red color was lost and turned less intense during thermal treatment. The *a*⋆ values were greater at lower heating temperatures (80 and 85 °C) which can be attributed to the anthocyanin contents. Anthocyanin and ascorbic acid interaction cause their degradation and the occurrence of nutritional and color changes (Choi, Kim, & Lee, 2002). A slight variation was observed in *b*⋆ value after 60 min of conventional or ohmic heating, which was attributed to carotenoid concentration in acerola. The results revealed that the carotenoid content was not affected by heating. The *L*⋆ values decreased during heating due to the ascorbic acid degradation and non-enzymatic browning reactions. Mercali et al. (2014) measured the effect of electric field frequency and heating time on color parameters of acerola and observed a decline over the time for all treatment conditions, which indicated color changes during heating. Higher color changes were observed at low electric field frequency (10 Hz), which might be due to the incidence of electrochemical reactions. The color changes in ohmic blanched artichokes were measured by Guida et al. (2013) and the results showed that the brightness (L^*) of ohmic blanched artichokes was enhanced by increasing the blanching time. However, L^* values were significantly decreased in conventionally blanched samples due to the incidence of browning reactions. Moreover, both blanching methods decreased hue angle and chroma and increased $\Delta E*$ values. The fresh-like color of ohmic-heated samples was due to the uniform and fast heating rate, which inactivated the enzymes in a short period of time. The color of artichoke was well preserved in ohmic heating, whereas hot water (conventional) blanching changed their color from yellow/green to brownish/green.

In addition to the effect of ohmic heating on the color of fruits and vegetables, several studies evaluated the effect of this processing on flavor. For instance, a better quality of strawberry jam in terms of color, flavor, and vitamin C content was reported using ohmic heating compared to microwave preservation technique, which was suggested to be due to the uniform and rapid heating and the absence of hot surfaces in ohmic heating apparatus (Avasoo & Johansson, 2011). Leizerson and Shimoni (2005a) investigated the impact of pasteurization of orange juice using ohmic heating in a very short time on its shelf life, which was compared to conventional pasteurization. In order to evaluate the quality of the obtained orange juice, the following parameters were studied: pectinesterase (PE) activity (i.e. cloudiness), ascorbic acid concentration, color, and flavor compounds (namely octanal, pinene, decanal, limonene, and myrcene). No significant difference was noticed between the two heating methods regarding PE activity and vitamin C concentration, although orange juice treated by ohmic heating appeared less cloudy. Similar effect on microbial load was observed for both heating methods during 105 days of storage. However, according to the sensory analysis, flavor in samples treated by ohmic heating was retained almost twice longer than the samples heated conventionally over a period of 100 days. The results showed that the characteristic flavor compounds were considerably higher in ohmic heated samples compared to conventional-treated juices (Leizerson & Shimoni, 2005b). Thanks to the presence of flavor compounds at similar levels, panelists were not able to differentiate between fresh and ohmic-treated juices. Similar results were observed by Anderson (2008) who demonstrated that the better flavor retention by ohmic heating was due to the shorter residence times and the absence of hot surfaces.

There are limited studies about the effects of ohmic cooking on other sensory attributes of plant-based products such as tenderness and crispiness. Hence, further studies need to be conducted to have a deeper understanding of the effect of this technique on the sensorial attributes of fruits and vegetables.

1.4.2 Meat Products

Sensory characteristics such as flavor, color, tenderness, and succulence have substantial influences on the acceptability of meat and meat products and can be affected by the cooking method. In ohmic processing, heat is distributed more rapidly and even throughout the product and result in improved flavor retention and structural integrity than the conventional heating methods. Shirsat et al. (2004) stated that the sensory evaluation of meat emulsion batters cooked by steam and ohmic heating had no perceivable difference. However, the study conducted by Zell et al. (2010b) exhibited that the color of ohmic-treated high-temperature short time (HTST) beef samples were more acceptable than low-temperature long time (LTLT) and conventionally cooked beef samples for the panelists. The ohmic heating offered a more uniform appearance, which was appreciated by the panelists. Moreover, the HTST treated samples were less juicy and firmer than the other products. These results were attributed to the higher cooking loss and end-point temperature in HTST samples. Accordingly, color and succulence of LTLT ohmic-cooked turkey meat were preferred to conventionally steam cooked samples, whereas the HTST treated samples were slightly firmer and less succulent. The conventional and HTST ohmic-heated turkey meat samples had a stronger odor intensity compared to LTLT-treated turkey meat samples (Zell et al., 2010a).

Generally, the color changes during cooking are due to myoglobin degradation through oxygenation, oxidation, and reduction reactions (Liu & Chen, 2001). Moreover, Maillard reaction has a significant role in surface color development in cooked meat products. In a study conducted by Engchuan et al. (2014), the influence of ohmic heating on color parameters of meatball was investigated. The hue angle and redness of conventional-cooked samples were higher than the ohmic-cooked counterparts. Similar results were reported by Bozkurt and Icier (2010) and Zell, Lyng, Cronin, and Morgan (2009) in ohmically and conventionally cooked beef muscles. They suggested that more browning reactions took place in the conventional-cooked meatballs compared to the ohmic-cooked samples. In the conventional method, the heat was mainly transferred by conduction and the outer layer of the meatballs would be much hotter than the core area. Consequently, browning occurred at the surface of conventional-cooked meatballs due to the high temperatures of the outer layer. However, in ohmic heating, the uniform temperature distribution inside the sample led to the formation of more homogeneous color and prevented the surface browning. Thus, the samples cooked by ohmic heating did not present a cooked crust, which is one of the disadvantages of ohmic cooking. Zell et al. (2009) also found that the samples heated at a slower heating rate had a slightly higher lightness and a^* and b^* values, which were lower than the samples heated at higher heating rates. These differences could be due to the dissimilarities of protein gel in meatball developed during fast or slow heating.

Dai et al. (2014) evaluated the color parameters of ohmic and water bath cooked pork and observed significantly lower a^* values but higher L^* and b^* values in conventionally cooked meat. Longer cooking time and higher exposure to oxygen in conventional cooking may induce increased oxidation of lipid and myoglobin pigments and inducing discolorations in conventionally-cooked meat (Fuentes et al., 2010; Ganhão, Morcuende, & Estévez, 2010). Shirsat et al. (2004) compared the color of ohmically and steam-cooked meat emulsion batter. Color assessments revealed that ohmic-cooked samples had lower hue angle and higher

$a*$ values compared to the steam cooked samples, while the panelists did not perceive the differences in the sensory evaluation.

Chai and Park (2007) determined the characteristics of fish proteins cooked with protein additives or starches under ohmic heating at 55 and 200 V. They found that surimi-starch gels that were ohmically cooked at 200 V were the most translucent samples after overnight refrigeration. This behavior might be due to the rapid heating time and reduced leaking of amylose chains into the gel system. Tadpitchayangkoon et al. (2012) reported substantial differences between ohmic and water-bath heated surimi gels. These changes could be attributed to rapid and short heating time in ohmic heating, which limited the browning reactions and preserved the whiteness of tropical surimi gels compared to water bath cooking. Moreover, the lightness depended on protein network structure and the differences in water-protein interactions under ohmic and water-bath heating resulted in changing the lightness of surimi gels.

Traditionally, meat emulsions are treated by water immersion or steam cooking. Lascorz et al. (2016) cooked shrimps by ohmic and steam cooking and found that all the color parameters (L^*, a^* and b^* values) of shrimp samples increased after cooking due to the release of astaxanthin from the carotenoproteins after denaturation. However, ohmic cooking of shrimps exhibited fewer color changes compared to steam cooking. This was suggested to be due to the short cooking time under ohmic heating.

1.4.3 Dairy Products

The flavor of milk is a key parameter of its quality and acceptance. Milk has a slightly salty and sweet taste owing to the presence of salts and lactose. Good quality milk has a bland but typical flavor with a nice mouthfeel (Thomas, 1981). Due to the bland flavor of milk, it can be the carrier of off-odors and flavors such as cooked, oxidized and rancid flavors (Azzara & Campbell, 1992). These unpleasant changes may occur during heating operations. The disruption of the milk fat globule membrane increases the accumulation of free fatty acids in the milk and leads to the formation of rancid flavors. Ohmic heating minimizes the generation of off-flavor compounds in milk and milk products because it does not promote more release of free fatty acids (FFA) (Ramaswamy, Marcotte, Sastry, & Abdelrahim, 2014). Irudayaraj et al. (2000) explored the impact of ohmic heating at different temperatures (135, 145, and 155 °C) and holding times (0.5 and 4.0 s) on UHT milk. They performed sensory evaluations and identified the volatile flavor compounds with gas chromatography (GC)-mass spectrometry (MS) to determine the influence of ohmic heating on organoleptic properties of UHT milk. They found significantly lower sour and stale flavors in the ohmic-heated samples compared to commercial UHT milk. In a study conducted by Roux et al. (2009), the color changes of infant formula during ohmic heating were measured. The lightness was decreased by temperature and heating time. However, increasing the temperature and heating time enhanced the $a*$ and $b*$ values, indicating that the redness and yellowness increased, which resulted in changing the color of the samples into brown. However, the results reported by Roux et al. (2016) revealed that the color of liquid infant formula remained unchanged with ohmic sterilization compared to that obtained after steam injection.

1.5 Advantageous and Challenges of Using Ohmic Processing on Food Quality

Application of ohmic processing has been reported to have several advantages over conventional treatment (Biss, Coombes, & Skudder, 1989; Khajehei et al., 2017). The foods objected to ohmic heating are uniformly heated in a well-controllable and rapid started process with the least thermal degradation. The lack of hot surface and its associated disadvantageous such as caramelization, fouling, and discoloration are among other attracting aspects. For foods in a container, the temperature in conventional heat processing must be high enough at the container walls to ensure sufficient heating of cold point (e.g. the center of big particulates) and guarantee the appropriate shelf life. This high-temperature can cause over processing and quality loss of the fluid (food) in contact with the walls. Due to the similar heating rate of particulates and the surrounding liquid, ohmic heating has overcome this limitation of conventional processing (Mojtahed Zadeh Asl, Niakousari, Hashemi Gahruie, Saharkhiz, & Mousavi Khaneghah, 2018). Because of the electrical resistance of the food components, heat is generated within the food as the electricity passes through it and both the liquid and solid parts will be warmed-up at the same time. The energy consumption is lower in the ohmic system and most of the energy is consumed at the control system and wires. The efficiency of ohmic heating is higher than 95% at most of the frequencies and using higher frequencies causes higher loss. Providing the high temperature using high-pressure steam (at a temperature of 200 °C, to increase the temperature of the product up to 180 °C, needs a pressure of 15 bar) in conventional processes is costly and limits its uses in the common processes. The high temperature (160–300 °C) used for sterilizing and partial hydrolysis of food waste before performing anaerobic digestion could be provided by ohmic heating. Flash depressurization is another way that can partially hydrolyze the particulates and so increase the volume of produced gas and also increase the bioavailability of nutrients after digestion.

The advantages of ohmic heating over conventional heating are summarized as:

- Rapid heating in short times for particulates and liquid, which can be usually achieved. This leads to protect vitamins and flavors compared with conventional heat exchangers.
- Low-temperature surfaces, resulting in a reduced risk of food damage from burning or over processing.

- Processing particulates or high solids' content. Low shear keeps particulate appearance and shape. Particulate products up to 3 cm are suitable for ohmic processing.
- High efficiency: 95% of the produced energy is transferred into the food.
- Highly controllable: suitable start-up and shutdown performance. Can be quickly joined with existing control systems.
- Robust/suitable resistance to fouling/good maintenance: automatic cleaning with standard clean in place operations effective due to minimal fouling of surfaces. Heaters can contain cooled surfaces for further mitigation into fouling with some foods.
- Suitable for foods with high viscosity.
- Suitable for high-temperature processing (e.g., working temperatures over 150 °C).
- Low-pressure and low shear drop all over the heaters.
- Can reduce the inefficiencies and infrastructure associated with the steam system.

The application of ohmic heating has several limitations that are summarized as:

- It is only suitable for foods with good electrical conductivity.
- Usually applied for those foods, which are pumpable and so ensure sufficiently suitable current transfer to and from ohmic electrodes.
- Designed units are mainly suited for a specific food conductivity range, while the design is much more difficult for extreme working conditions.
- The electric nature of this heating system means that the sites considering ohmic systems will need a good electrical supply, and the electrical energy cost compared with fuels must be taken into account when considering the ohmic system.

1.6 Conclusions

The results of these studies showed that ohmic heating is faster, less aggressive, and may improve the overall quality of products compared to the conventional heating systems. The high quality in ohmic-heated food products is attributed to the uniform heat generation throughout the product, which minimizes the overcooked areas in foods and preserves their texture, color, taste, and nutrients such as fatty acids, vitamins, amino acids, and phenolic compounds. Hence ohmic heating can be an appropriate alternative to conventional heating allowing food manufacturers to acquire high-quality products. However, the process variables such as temperature, voltage gradients, electric field frequency, and heating time play a dominant role in the quality of ohmic-heated samples. Therefore, these parameters should be selected cautiously to reach the desired quality in ohmic treated food products.

References

Aamir, M., & Jittanit, W. (2017). Ohmic heating treatment for Gac aril oil extraction: Effects on extraction efficiency, physical properties and some bioactive compounds. *Innovative Food Science & Emerging Technologies, 41*, 224–234.

Achir, N., Dhuique-Mayer, C., Hadjal, T., Madani, K., Pain, J.-P., & Dornier, M. (2016). Pasteurization of citrus juices with ohmic heating to preserve the carotenoid profile. *Innovative Food Science & Emerging Technologies, 33*, 397–404.

Anderson, A. K., & Finkelstein, R. (1919). A study of the electro-pure process of treating milk. *Journal of Dairy Science, 2*(5), 374–406.

Anderson, D. R. (2008). *Ohmic heating as an alternative food processing technology* (Master's thesis). Kansas state University, Kansas, USA.

Athmaselvi, K. A., Kumar, C., & Poojitha, P. (2017). Influence of temperature, voltage gradient and electrode on ascorbic acid degradation kinetics during ohmic heating of tropical fruit pulp. *Journal of Food Measurement and Characterization, 11*(1), 144–155.

Avasoo, M., & Johansson, L. (2011). *Evaluation of thermal processing technologies for strawberry jam* (Master's thesis in food technology). Sweden: Lund University.

Azzara, C. D., & Campbell, L. B. (1992). Off-flavors of dairy products. In G. Charalambous (Ed.), *Developments in food science* (Vol. 28, pp. 329–374). Amsterdam, The Netherlands: Elsevier Science Publishers.

Bastías, J. M., Moreno, J., Pia, C., Reyes, J., Quevedo, R., & Muñoz, O. (2015). Effect of ohmic heating on texture, microbial load, and cadmium and lead content of Chilean blue mussel (*Mytilus chilensis*). *Innovative Food Science & Emerging Technologies, 30*, 98–102.

Biss, C., Coombes, S., & Skudder, P. (1989). Process engineering in the food industry: Developments and opportunities. In R. W. Field & J. A. Howell (Eds.), *Process engineering in the food industry: Developments and opportunities*. Essex, EN: Elsevier Applied Science.

Bourne, M. C. (2002). *Food texture and viscosity: Concept and measurement*. San Diego, CA: Academic.

Bozkurt, H., & Icier, F. (2010). Ohmic cooking of ground beef: Effects on quality. *Journal of Food Engineering, 96*(4), 481–490.

Castro, I., Macedo, B., Teixeira, J. A., & Vicente, A. A. (2004). The effect of electric field on important food-processing enzymes: Comparison of inactivation kinetics under conventional and ohmic heating. *Journal of Food Science, 69*(9), C696–C701.

Castro, I., Teixeira, J. A., Salengke, S., Sastry, S. K., & Vicente, A. A. (2004). Ohmic heating of strawberry products: Electrical conductivity measurements and ascorbic acid degradation kinetics. *Innovative Food Science & Emerging Technologies, 5*(1), 27–36.

Chai, P. P., & Park, J. W. (2007). Physical properties of fish proteins cooked with starches or protein additives under ohmic heating. *Journal of Food Quality, 30*(5), 783–796.

Chakraborty, I., & Athmaselvi, K. A. (2014). Changes in physicochemical properties of guava juice during ohmic heating. *Journal of Ready to Eat Food, 1*(4), 152–157.

Chiavaro, E., Barbanti, D., Vittadini, E., & Massini, R. (2006). The effect of different cooking methods on the instrumental quality of potatoes (cv. Agata). *Journal of Food Engineering, 77*(1), 169–178.

Choi, M. H., Kim, G. H., & Lee, H. S. (2002). Effects of ascorbic acid retention on juice color and pigment stability in blood orange (*Citrus sinensis*) juice during refrigerated storage. *Food Research International, 35*(8), 753–759.

Christian, G., & Leadley, C. (2006). *New technologies bulletin 32* (No. GL55 6LD). UK: Campden & Chorleywood Food Research Association, Chipping Campden, Gloucestershire.

Dai, Y., Lu, Y., Wu, W., Lu, X., Han, Z., Liu, Y., … Dai, R. (2014). Changes in oxidation, color and texture deteriorations during refrigerated storage of ohmically and water bath-cooked pork meat. *Innovative Food Science & Emerging Technologies, 26*, 341–346.

Damodaran, S., Parkin, K., & Fennema, O. R. (2008). *Fennema's food chemistry*. Boca Raton, FL: CRC Press.

Damyeh, M. S., Niakousari, M., Golmakani, M. T., & Saharkhiz, M. J. (2016). Microwave and ohmic heating impact on the in situ hydrodistillation and selective extraction of *Satureja macrosiphonia* essential oil. *Journal of Food Processing and Preservation, 40*(4), 647–656.

Damyeh, M. S., Niakousari, M., Saharkhiz, M. J., & Golmakani, M. T. (2016). Evaluating the effect of essential oil extraction method from *Satureja macrosiphonia* on its biological activities: Ohmic-and microwave-assisted hydrodistillation. *Journal of Food Processing and Preservation, 40*(4), 697–706.

Dima, F., Istrati, D., Garnai, M., Serea, V., & Vizireanu, C. (2015). Study on obtaining vegetables juices with high antioxidant potential, preserved by ohmic pasteurization. *Journal of Agroalimentary Processes and Technologies, 21*, 67–74.

Duygu, B., & Ümit, G. (2015). Application of ohmic heating system in meat thawing. *Procedia - Social and Behavioral Sciences, 195*, 2822–2828.

Engchuan, W., Jittanit, W., & Garnjanagoonchorn, W. (2014). The ohmic heating of meat ball: Modeling and quality determination. *Innovative Food Science & Emerging Technologies, 23*, 121–130.

Farahnaky, A., Azizi, R., & Gavahian, M. (2012). Accelerated texture softening of some root vegetables by Ohmic heating. *Journal of Food Engineering, 113*(2), 275–280.

Foegeding, E. A., Allen, C. E., & Dayton, W. R. (1986). Effect of heating rate on thermally formed myosin, fibrinogen and albumin gels. *Journal of Food Science, 51*(1), 104–108.

Fuentes, V., Ventanas, J., Morcuende, D., Estévez, M., & Ventanas, S. (2010). Lipid and protein oxidation and sensory properties of vacuum-packaged dry-cured ham subjected to high hydrostatic pressure. *Meat Science, 85*(3), 506–514.

Ganhão, R., Morcuende, D., & Estévez, M. (2010). Protein oxidation in emulsified cooked burger patties with added fruit extracts: Influence on colour and texture deterioration during chill storage. *Meat Science, 85*(3), 402–409.

Gatellier, P., Kondjoyan, A., Portanguen, S., & Santé-Lhoutellier, V. (2010). Effect of cooking on protein oxidation in n-3 polyunsaturated fatty acids enriched beef. Implication on nutritional quality. *Meat Science, 85*(4), 645–650.

Guida, V., Ferrari, G., Pataro, G., Chambery, A., Di Maro, A., & Parente, A. (2013). The effects of ohmic and conventional blanching on the nutritional, bioactive compounds and quality parameters of artichoke heads. *LWT - Food Science and Technology, 53*(2), 569–579.

Hashemi Gahruie, H., Hosseini, S. M. H., Taghavifard, M. H., Eskandari, M. H., Golmakani, M.-T., & Shad, E. (2017). Lipid oxidation, color changes, and microbiological quality of frozen beef burgers incorporated with shirazi thyme, cinnamon, and rosemary extracts. *Journal of Food Quality*. https://doi.org/10.1155/2017/6350156

Hradecky, J., Kludska, E., Belkova, B., Wagner, M., & Hajslova, J. (2017). Ohmic heating: A promising technology to reduce furan formation in sterilized vegetable and vegetable/meat baby foods. *Innovative Food Science and Emerging Technologies, 43*, 1–6.

Hurtaud, C., & Peyraud, J. L. (2007). Effects of feeding camelina (seeds or meal) on milk fatty acid composition and butter spreadability. *Journal of Dairy Science, 90*(11), 5134–5145.

Icier, F. (2009). Influence of ohmic heating on rheological and electrical properties of reconstituted whey solutions. *Food and Bioproducts Processing, 87*(4), 308–316.

Icier, F., Izzetoglu, G. T., Bozkurt, H., & Ober, A. (2010). Effects of ohmic thawing on histological and textural properties of beef cuts. *Journal of Food Engineering, 99*(3), 360–365.

Irudayaraj, J., McMahon, D., & Reznik, D. (2000). *Ohmic heating for UHT milk*. Presented at the Annual Meeting, Institute of Food Technologists, Dallas, Texas.

Jaeschke, D. P., Marczak, L. D. F., & Mercali, G. D. (2016). Evaluation of non-thermal effects of electricity on ascorbic acid and carotenoid degradation in acerola pulp during ohmic heating. *Food Chemistry, 199*, 128–134.

Jittanit, W., Khuenpet, K., Kaewsri, P., Dumrongpongpaiboon, N., Hayamin, P., & Jantarangsri, K. (2017). Ohmic heating for cooking rice: Electrical conductivity measurements, textural quality determination and energy analysis. *Innovative Food Science & Emerging Technologies, 42*, 16–24.

Kamali, E., & Farahnaky, A. (2015). Ohmic-assisted texture softening of cabbage, turnip, potato and radish in comparison with microwave and conventional heating. *Journal of Texture Studies, 46*(1), 12–21.

Khajehei, F., Niakousari, M., Seidi Damyeh, M., Merkt, N., Claupein, W., & Graeff-Hoenninger, S. (2017). Impact of ohmic-assisted decoction on bioactive components extracted from yacon (*Smallanthus sonchifolius* Poepp.) leaves: Comparison with conventional decoction. *Molecules, 22*(12). https://doi.org/10.3390/molecules22122043

Kim, J.-Y., Hong, G.-P., Park, S.-H., Spiess, W. E., & Min, S.-G. (2006). Effect of ohmic thawing on physico-chemical properties of frozen hamburger patties. *Korean Journal for Food Science of Animal Resources, 26*(2), 223–228.

Kim, N. H., Ryang, J. H., Lee, B. S., Kim, C. T., & Rhee, M. S. (2017). Continuous ohmic heating of commercially processed apple juice using five sequential electric fields results in rapid inactivation of *Alicyclobacillus acidoterrestris* spores. *International Journal of Food Microbiology, 246*, 80–84.

Kim, Y. H., Huff-Lonergan, E., Sebranek, J. G., & Lonergan, S. M. (2010). High-oxygen modified atmosphere packaging system induces lipid and myoglobin oxidation and protein polymerization. *Meat Science, 85*(4), 759–767.

Koubaa, M., Roselló-Soto, E., Barba-Orellana, S., & Barba, F. J. (2016). Novel thermal technologies and fermentation. In K. S. Ojha & B. K. Tiwari (Eds.), *Novel food fermentation technologies* (pp. 155–163). Cham, Switzerland: Springer.

Lascorz, D., Torella, E., Lyng, J. G., & Arroyo, C. (2016). The potential of ohmic heating as an alternative to steam for heat processing shrimps. *Innovative Food Science & Emerging Technologies, 37*, 329–335.

Leizerson, S., & Shimoni, E. (2005a). Effect of ultrahigh-temperature continuous ohmic heating treatment on fresh orange juice. *Journal of Agricultural and Food Chemistry, 53*(9), 3519–3524.

Leizerson, S., & Shimoni, E. (2005b). Stability and sensory shelf life of orange juice pasteurized by continuous ohmic heating. *Journal of Agricultural and Food Chemistry, 53*(10), 4012–4018.

Liu, Y., & Chen, Y. R. (2001). Analysis of visible reflectance spectra of stored, cooked and diseased chicken meats. *Meat Science, 58*(4), 395–401.

Louarme, L., & Billaud, C. (2012). Evaluation of ascorbic acid and sugar degradation products during fruit dessert processing under conventional or ohmic heating treatment. *LWT - Food Science and Technology, 49*(2), 184–187.

Lund, M. N., Lametsch, R., Hviid, M. S., Jensen, O. N., & Skibsted, L. H. (2007). High-oxygen packaging atmosphere influences protein oxidation and tenderness of porcine longissimus dorsi during chill storage. *Meat Science, 77*(3), 295–303.

Lutz, M., Henríquez, C., & Escobar, M. (2011). Chemical composition and antioxidant properties of mature and baby artichokes (*Cynara scolymus* L.), raw and cooked. *Journal of Food Composition and Analysis, 24*(1), 49–54.

Mathoniere, C., Mioche, L., Dransfield, E., & Culioli, J. (2000). Meat texture characterisation: Comparison of chewing patterns, sensory and mechanical measures. *Journal of Texture Studies, 31*(2), 183–203.

Mercali, G. D., Gurak, P. D., Schmitz, F., & Marczak, L. D. F. (2015). Evaluation of non-thermal effects of electricity on anthocyanin degradation during ohmic heating of jaboticaba (*Myrciaria cauliflora*) juice. *Food Chemistry, 171*, 200–205.

Mercali, G. D., Jaeschke, D. P., Tessaro, I. C., & Marczak, L. D. F. (2012). Study of vitamin C degradation in acerola pulp during ohmic and conventional heat treatment. *LWT – Food Science and Technology, 47*(1), 91–95.

Mercali, G. D., Schwartz, S., Marczak, L. D. F., Tessaro, I. C., & Sastry, S. (2014). Ascorbic acid degradation and color changes in acerola pulp during ohmic heating: Effect of electric field frequency. *Journal of Food Engineering, 123*, 1–7.

Mesías, M., Wagner, M., George, S., & Morales, F. J. (2016). Impact of conventional sterilization and ohmic heating on the amino acid profile in vegetable baby foods. *Innovative Food Science & Emerging Technologies, 34*, 24–28.

Mojtahed Zadeh Asl, R., Niakousari, M., Hashemi Gahruie, H., Saharkhiz, M. J., & Mousavi Khaneghah, A. (2018). Study of two-stage ohmic hydro-extraction of essential oil from *Artemisia aucheri* Boiss.: Antioxidant and antimicrobial characteristics. *Food Research International, 107*, 462–469.

Moreno, J., Echeverria, J., Silva, A., Escudero, A., Petzold, G., Mella, K., & Escudero, C. (2017). Apple snack enriched with L-arginine using vacuum impregnation/ohmic heating technology. *Food Science and Technology International, 23*(5), 448–456.

Niakousari, M., Hashemi Gahruie, H., Razmjooei, M., Roohinejad, S., & Greiner, R. (2018). Effects of innovative processing technologies on microbial targets based on food categories: Comparing traditional and emerging technologies for food preservation. In F. J. Barba, A. S. Sant'Ana, V. Orlien, & M. Koubaa (Eds.), *Innovative technologies for food preservation: Inactivation of spoilage and pathogenic microorganisms* (pp. 133–185). London, UK: Academic Press, Elsevier.

Oey, I., Lille, M., Van Loey, A., & Hendrickx, M. (2008). Effect of high-pressure processing on colour, texture and flavour of fruit- and vegetable-based food products: A review. *Trends in Food Science & Technology, 19*(6), 320–328.

Olivera, D. F., Salvadori, V. O., & Marra, F. (2013). Ohmic treatment of fresh foods: Effect on textural properties. *International Food Research Journal, 20*(4), 1617–1621.

Özkan, N., Ho, I., & Farid, M. (2004). Combined ohmic and plate heating of hamburger patties: Quality of cooked patties. *Journal of Food Engineering, 63*(2), 141–145.

Parmar, P. R., Tripathi, S., Tiwari, S., & Singh, R. (2016). Fabrication and performance evaluation of ohmic heater. *International Journal of Research in Science and Technology, 6*(4), 59–69.

Patyukov, S., & Pacinovski, N. (2015). Effect of traditional and ohmic heating on fat stability of pufa-fortified cooked sausages. *Macedonian Journal of Animal Science, 5*(2), 107–112.

Pedersen, S. J., Feyissa, A. H., Brøkner Kavli, S. T., & Frosch, S. (2016). An investigation on the application of ohmic heating of cold water shrimp and brine mixtures. *Journal of Food Engineering, 179*, 28–35.

Pereira, R. N., Martins, R. C., & Vicente, A. A. (2008). Goat milk free fatty acid characterization during conventional and ohmic heating pasteurization. *Journal of Dairy Science, 91*(8), 2925–2937.

Pereira, R. N., Rodrigues, R. M., Ramos, Ó. L., Xavier Malcata, F., Teixeira, J. A., & Vicente, A. A. (2016). Production of whey protein-based aggregates under ohmic heating. *Food and Bioprocess Technology, 9*(4), 576–587.

Purchas, R. W., Wilkinson, B. H. P., Carruthers, F., & Jackson, F. (2014). A comparison of the nutrient content of uncooked and cooked lean from New Zealand beef and lamb. *Journal of Food Composition and Analysis, 35*(2), 75–82.

Rahman, S. (2007). *Handbook of food preservation*. Boca Raton, FL: CRC Press.

Ramaswamy, H. S., Marcotte, M., Sastry, S., & Abdelrahim, K. (2014). *Ohmic heating in food processing*. Boca Raton, FL: CRC Press.

Roberts, J. S., Balaban, M. O., Zimmerman, R., & Luzuriaga, D. (1998). Design and testing of a prototype ohmic thawing unit. *Computers and Electronics in Agriculture, 19*(2), 211–222.

Rodriguez-Amaya, D. B. (2001). *A guide to carotenoid analysis in foods (ILSI press)*. Washington, D.C.: ILSI Press.

Roux, S., Courel, M., Ait-Ameur, L., Birlouez-Aragon, I., & Pain, J.-P. (2009). Kinetics of Maillard reactions in model infant formula during UHT treatment using a static batch ohmic heater. *Dairy Science & Technology, 89*(3), 349–362.

Roux, S., Courel, M., Birlouez-Aragon, I., Municino, F., Massa, M., & Pain, J.-P. (2016). Comparative thermal impact of two UHT technologies, continuous ohmic heating and direct steam injection, on the nutritional properties of liquid infant formula. *Journal of Food Engineering, 179*, 36–43.

Santé-Lhoutellier, V., Astruc, T., Marinova, P., Greve, E., & Gatellier, P. (2008). Effect of meat cooking on physicochemical state and in vitro digestibility of myofibrillar proteins. *Journal of Agricultural and Food Chemistry, 56*(4), 1488–1494.

Sarkis, J. R., Jaeschke, D. P., Tessaro, I. C., & Marczak, L. D. F. (2013). Effects of ohmic and conventional heating on anthocyanin degradation during the processing of blueberry pulp. *LWT – Food Science and Technology, 51*(1), 79–85.
Seidi Damyeh, M., & Niakousari, M. (2016). Impact of ohmic-assisted hydrodistillation on kinetics data, physicochemical and biological properties of *Prangos ferulacea* Lindle. essential oil: Comparison with conventional hydrodistillation. *Innovative Food Science & Emerging Technologies, 33*, 387–396.
Seidi Damyeh, M., & Niakousari, M. (2017). Ohmic hydrodistillation, an accelerated energy-saver green process in the extraction of *Pulicaria undulata* essential oil. *Industrial Crops and Products, 98*, 100–107.
Septembre-Malaterre, A., Remize, F., & Poucheret, P. (2018). Fruits and vegetables, as a source of nutritional compounds and phytochemicals: Changes in bioactive compounds during lactic fermentation. *Food Research International, 104*, 86–99.
Shirsat, N., Brunton, N. P., Lyng, J. G., McKenna, B., & Scannell, A. (2004). Texture, colour and sensory evaluation of a conventionally and ohmically cooked meat emulsion batter. *Journal of the Science of Food and Agriculture, 84*(14), 1861–1870.
Shynkaryk, M. V., Ji, T., Alvarez, V. B., & Sastry, S. K. (2010). Ohmic heating of peaches in the wide range of frequencies (50 Hz to 1 MHz). *Journal of Food Science, 75*(7), E493–E500.
Simpson, D. P. (1983). *Apparatus for heating electrically conductive flowable media*. Patent number: DE3160372D1.
Simpson, D. P., & Stirling, R. (1995). *Ohmic Heater including electrodes arranged along a flow axis to reduce leakage current*. Patent number: US5440667A.
Singh, A., Santosh, S., Kulshrestha, M., Chand, K., Lohani, U., & Shahi, N. (2013). Quality characteristics of Ohmic heated Aonla (*Emblica officinalis* Gaertn.) pulp. *Indian Journal of Traditional Knowledge, 12*(4), 670–676.
Stirling, R., & Coombes, S. A. (1990). *Apparatus for heating an electrically conductive flowable material flowing through a pipeline*. Patent number: US4959525A.
Tadpitchayangkoon, P., Park, J. W., & Yongsawatdigul, J. (2012). Gelation characteristics of tropical surimi under water bath and ohmic heating. *LWT – Food Science and Technology, 46*(1), 97–103.
Thomas, E. L. (1981). Trends in milk flavors. *Journal of Dairy Science, 64*(6), 1023–1027.
Toldrá, F. (2017). *Lawrie's meat science*. Duxford, UK: Woodhead Publishing.
Tumpanuvatr, T., & Jittanit, W. (2012). The temperature prediction of some botanical beverages, concentrated juices and purees of orange and pineapple during ohmic heating. *Journal of Food Engineering, 113*(2), 226–233.
Van Buren, J. P. (1979). The chemistry of texture in fruits and vegetables. *Journal of Texture Studies, 10*(1), 1–23.
Vikram, V. B., Ramesh, M. N., & Prapulla, S. G. (2005). Thermal degradation kinetics of nutrients in orange juice heated by electromagnetic and conventional methods. *Journal of Food Engineering, 69*(1), 31–40.
Waldron, K. W., Parker, M. L., & Smith, A. C. (2003). Plant cell walls and food quality. *Comprehensive Reviews in Food Science and Food Safety, 2*(4), 128–146.
Wells, A. S. (2001). The role of milk in the British diet. *International Journal of Dairy Technology, 54*(4), 130–134.
Wills, T. M., Dewitt, C. A. M., Sigfusson, H., & Bellmer, D. (2006). Effect of cooking method and ethanolic tocopherol on oxidative stability and quality of beef patties during refrigerated storage (oxidative stability of cooked patties). *Journal of Food Science, 71*(3), C109–C114.
Zell, M., Lyng, J. G., Cronin, D. A., & Morgan, D. J. (2009). Ohmic cooking of whole beef muscle – Optimisation of meat preparation. *Meat Science, 81*(4), 693–698.
Zell, M., Lyng, J. G., Cronin, D. A., & Morgan, D. J. (2010a). Ohmic cooking of whole beef muscle-evaluation of the impact of a novel rapid ohmic cooking method on product quality. *Meat Science, 86*(2), 258–263.
Zell, M., Lyng, J. G., Cronin, D. A., & Morgan, D. J. (2010b). Ohmic cooking of whole turkey meat – Effect of rapid ohmic heating on selected product parameters. *Food Chemistry, 120*(3), 724–729.

Chapter 2
Effects of Pulsed Electric Fields on Food Constituents, Microstructure and Sensorial Attributes of Food Products

Shualing Yang, Guanchen Liu, Zihan Qin, Daniel Munk, Jeanette Otte, and Lilia Ahrné

2.1 Introduction

To assess the tangible advantages of pulsed electric fields (PEF) as alternative preservation technology, or to explore its use for improvement of functionality and healthiness of foods and ingredients or creating novel food structures, knowledge is needed regarding their impact on key food constituents such as proteins, lipids, carbohydrates, bioactive and flavor compounds, as well as on product microstructure. The successful application of the technology requires also that sensorial attributes of the PEF-treated food meet the consumer expectations and acceptance. In this chapter, studies on the impact of PEF on these key issues will be discussed, in both animal- and plant-based foods.

2.2 On Food Constituents

2.2.1 Proteins

Proteins are important nutritional and functional components of foods. Furthermore, they are polyelectrolytes, which are easily affected by chemical and physical treatments. Therefore, a number of studies have been done to evaluate the effects of PEF on various food proteins and consequences for functionality.

S. Yang · G. Liu · Z. Qin · D. Munk · J. Otte · L. Ahrné (✉)
Department of Food Science, University of Copenhagen, Copenhagen, Denmark
e-mail: lilia@food.ku.dk

S. Roohinejad et al. (eds.), *Effect of Emerging Processing Methods on the Food Quality*,
https://doi.org/10.1007/978-3-030-18191-8_2

2.2.1.1 Milk Proteins

Caseins and whey proteins constitute the major proteins of milk. Caseins consist of four different proteins, α_{s1}-, α_{s2}-, β-, and κ-casein, which are hydrophobic, phosphorylated proteins, and occur in milk as large colloidal structures, so-called casein micelles (Dalgleish, 2011; Walstra, Wouters, & Geurts, 2014). Each casein micelle consists of several thousands of individual casein protein molecules and the size varies from 80 to 400 nm, with an average size of about 200 nm (Dalgleish, 2011; McMahon & Oommen, 2013). The structure of the casein micelles has not been fully elucidated, but there is general agreement that they have a rather open porous structure containing a large amount of water molecules (McMahon & Oommen, 2013), and also that the α_s-casein and β-casein molecules are located primarily in the interior of the casein micelles, while κ-casein, which is negatively charged, has glycosylated C-terminal and only few phosphorylation sites, is located at the exterior with the C-terminal part protruding out from the micelle into the surrounding medium. A recent model of a casein micelle is shown in Fig. 2.1. The caseins are held together in the micelle by both hydrophobic interactions between hydrophobic parts of the casein molecules and colloidal calcium phosphate interactions between phosphorylated parts of the casein molecules (Dalgleish, 2011; Horne, 1998; McMahon & Oommen, 2013; Walstra, 1990). Therefore, they are sensitive to treatments (e.g. temperature, pH, and pressure) that affect these kinds of protein-protein interactions and the distribution of calcium in milk.

The studies published regarding the effects of PEF treatment on the caseins in milk are presented in the upper part of Table 2.1. All studies have been performed with skimmed milk and no study was found about the effect of PEF on caseins in whole milk or casein ingredients. The PEF treatment seems to have an effect on the size of casein micelles but inconsistent results have been reported. Floury et al. (2006) showed that PEF processing at field levels of 45–55 kV/cm at a temperature

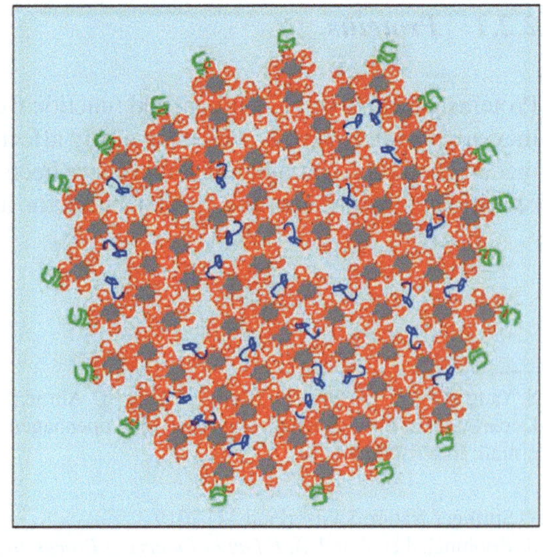

Fig. 2.1 Schematic structure of the casein micelle, incorporating calcium phosphate nanoclusters (grey) with their attached casein (red) and the surface-located κ-casein (green). The "hydrophobically bound" mobile β-casein is shown in blue, within the water channels inside the micelle. (Reproduced with permission from Dalgleish (2011))

Table 2.1 Effect of pulsed electric field treatments on dairy proteins

Protein	Product	Treatment intensity	Effects	Reference
Caseins	Skim milk	45–55 kV/cm for 2.1–3.5 μs; T < 50 °C	Reduced size of casein micelles	Floury et al. (2006)
		45 kV/cm for 20 μs; T_i = 25 °C; T_o = 30 °C	No decrease in size of casein micelles	Hemar et al. (2011)
		35 kV/cm for 188 μs; T_i = 22 °C; T_o = 52 °C	No decrease in size of casein micelles	Michalac, Alvarez, Ji, and Zhang (2003)
		49 kV/cm for 20 μs; T ≤ 70 °C	Reduced size of casein micelles at pH 8.0	Liu et al. (2015)
Whey proteins	β-LG solution	30 kV/cm for 1.3 μs; T ≤ 30 °C	No significant unfolding or aggregation of β-LG	Barsotti, Dumay, Mu, Fernandez Diaz, and Cheftel (2001)
	IgG solution / enriched soy milk	41.1 kV/cm for 54 μs; T < 50 °C; T_i = 15 °C; T_o = 43.8 °C	No effects on secondary structure of IgG	Li, Zhang, Lee, and Pham (2003)
	WPI solution	30 kV/cm for 19.2 and 211 μs; T = 30, 60, 65, 70, or 75 °C	No effect on physicochemical properties of whey proteins	Sui, Roginski, Williams, Versteeg, and Wan (2011)
	Whole milk	35.5 kV/cm for 1000 μs; T ≤ 40 °C	β-LG, α-LA, and BSA partially denatured (25–40%)	Odriozola-Serrano, Bendicho-Porta, and Martín-Belloso (2006)
	WPI solution	12–20 kV/cm, 10–20 pulses; T_o < 35 °C	Partial denaturation of WPI fractions	Xiang, Ngadi, Ochoa-Martinez, and Simpson (2011)
	β-LG concentrate	12.5 kV/cm for 2000 μs; T < 35 °C	Increased gelation properties of β-LG	Perez and Pilosof (2004)

below 50 °C (Europulse, Cressensac, France) with 2.1–3.5 μs treatment time reduced the size of casein micelles significantly in skim milk. However, no decrease in casein micelle size at a similar field level (45 kV/cm), but kept for longer time (20 μs up to 30 °C), has been reported by Hemar et al. (2011) and (35 kV/cm with 188 μs up to 52 °C) by Michalac et al. (2003). Liu et al. (2015) also found that PEF treatment (49 kV/cm for about 20 μs up to 70 °C) did not change the size of the casein micelles in milk at the natural pH but reduced the size of the casein micelles at higher pH (pH 8). The varying results are probably due to different types of PEF equipment used, and different pre-treatments of milk before PEF (e.g. defatting and pasteurization).

The major whey proteins (WPs) in bovine milk are β-lactoglobulin (β-LG) and α-lactalbumin (α-LA), followed by bovine serum albumin (BSA), immunoglobulins (Ig), lactoferrin, and various other minor proteins and enzymes (Walstra et al., 2014). WPs, unlike caseins, are globular proteins with a compact tertiary structure held together by disulfide bonds. The major WP, β-LG, furthermore, contains a free thiol group, which is buried in the hydrophobic interior of the native molecule. Upon partial denaturation of β-LG, this group will be exposed together with other hydrophobic parts of the molecule and can engage in thiol/disulfide interchange and other reactions with other milk proteins (e.g. WPs to form WP aggregates).

The studies regarding the effects of PEF on whey proteins in model single protein systems or milk are shown in the lower part of Table 2.1. Some studies reported that PEF processing does not cause notable protein unfolding or aggregation of β-LG and does not affect the concentration and antigen binding activity of immunoglobulin G (IgG) in solution (Barsotti et al., 2001; Li et al., 2003; Li, Bomser, & Zhang, 2005). Barsotti et al. (2001) applied PEF (30 kV/cm for 1.3 μs) to β-LG solutions in sodium phosphate buffer (pH 7) and did not observe significant unfolding or aggregation of β-LG. Similarly, Li et al. (2003) found no effects of PEF processing (41.1 kV/cm for 54 μs up to 43.8 °C) on the secondary structure of IgG in enriched soy milk. The recent study by Sui et al. (2011), likewise, showed that PEF processing (30 kV/cm for 19.2 and 211 μs up to 75 °C) of whey protein isolate (WPI) in simulated milk ultrafiltrate (pH 6.5) did not change the physicochemical properties of the WPs, such as protein aggregation, surface hydrophobicity, contents of exposed and total sulphydryl groups and thermal stability. However, in whole milk and for longer treatment time (1000 μs), Odriozola-Serrano et al. (2006) reported that β-LG, α-LA, and BSA were partially denatured (25–40%) after PEF treatment at 35.5 kV/cm. Similarly, Xiang, Ngadi, et al. (2011) found that PEF treatment (12–20 kV/cm, 10–20 pulses, up to 35 °C) of WPI solutions in distilled water (pH 7) resulted in partial denaturation of WPI fractions and exposure of more hydrophobic regions. This is in line with the improved gelation properties of a β-LG concentrate (10% in water, pH 7) observed after PEF treatment at 12.5 kV/cm for a long treatment time (up to 2000 μs) (Perez & Pilosof, 2004).

The use of PEF-treated milk to produce cheese has been studied by Sepúlveda-Ahumada, Ortega-Rivas, and Barbosa-Cánovas (2000). The results showed no significant differences in the structure of milk proteins or texture of Cheddar cheese produced with PEF treated milk and that produced with HTST pasteurized milk.

The studies reported so far, thus, show that PEF processing for a long time can affect the whey proteins and expose the hydrophobic and thiol groups that were originally buried in the proteins, leading to whey protein aggregation and gelation through hydrophobic and thiol/disulfide interactions. However, it cannot be ruled out that these changes may be (partly) due to a temperature increase associated with the long PEF treatment time. More fundamental studies are needed to understand the protein modification due to PEF treatment alone.

2.2.1.2 Egg Proteins

The major egg white proteins are ovalbumin (54%), conalbumin (12%), ovomucoid (11%), lysozyme (3.5%), and ovomucin (2–4%) (Mine, 2015). Ovalbumin is a phosphoglycoprotein that can be converted to s-ovalbumin during cold storage. It is the most heat stable egg protein with a denaturation temperature of 84 °C, but it can be easily denatured and forms gels by physical treatments such as heating and high hydrostatic pressure processing. Conalbumin (also called ovotransferrin), ovomucoid, and ovomucin are glycoproteins. Conalbumin has the capacity to bind bi- and trivalent metal cations into a complex, making the protein more heat stable than in the native state. Ovomucoid occurs in various forms, which differ in the amount of carbohydrate bonded via asparagine. Nine disulfide bridges and a high content (<80%) of helical and β-pleated sheet structures confer upon ovomucoid such a stable spatial structure that it is not denatured even upon boiling. This protein is the major allergen found in egg and was found to be a trypsin inhibitor. Ovomucin is a relatively small protein (MW = 10 kDa) with a diversity of carbohydrate bound and a large portion of hydroxyl groups esterified with sulfuric acid. The ovomucin aggregates into filamentous and fibroid structures being responsible for the high viscosity of the egg white. Lysozyme is an enzyme that can lyse the wall of certain Gram-positive bacteria and is found at high levels in the chalaziferous layer and the chalazae, which anchor the yolk towards the middle of the egg.

Sampedro et al. (2006) reviewed the application of pulsed electric fields in egg and egg products, concluding that the high electric conductivity and the complex composition of whole eggs, having high protein and fat contents, is a challenge for the application of PEF as preservation technology. This high protein concentration may also be the reason why studies about the effects of PEF on egg white, have shown that PEF significantly influences the thiol group reactivity of egg proteins, depending on the number of pulses and the energy applied per pulse (Table 2.2). The egg white proteins were partially denatured and altered in terms of surface hydrophobicity, exposed SH groups and gelation properties, as well as, emulsifying capacity and stability in fresh egg white after PEF treatment at 12.5 kV/cm for 2300 μs up to 35 °C, and 25–35 kV/cm for 400–800 μs up to 42 °C (Perez & Pilosof, 2004; Zhao et al., 2007). However, the changes in surface hydrophobicity of egg white proteins have not been observed in earlier studies by Jeantet et al. (1999), who

Table 2.2 Effect of pulsed electric field treatments on egg proteins

Protein	Product	Treatment intensity	Effects	Reference
Egg proteins	Egg white	12.5 kV/cm for 2300 μs; T < 35 °C	Partially denatured egg white proteins	Perez and Pilosof (2004)
		25–35 kV/cm for 400–800 μs; T = 40 °C	Partially denatured egg white proteins	Zhao, Yang, Tang, and Lu (2007)
		20–35 kV/cm, 2–8 pulses; T = 4–30 °C	No increase in surface hydrophobicity of egg white proteins	Jeantet, Baron, Nau, Roignant, and Brulé (1999)
	Liquid whole egg	19 and 32 kV/cm for 30 μs and 37 kV/cm for 18 μs	No significant differences in the water-soluble protein fraction	Marco-Molés et al. (2011)
	Ovalbumin solution	27–33 kV/cm for 0.3–0.9 μs; T < 29 °C	No permanent modification of ovalbumin	Fernandez-Diaz, Barsotti, Dumay, and Cheftel (2000)

found no change by applying 2–8 exponential pulses of 20–35 kV/cm. Fernandez-Diaz et al. (2000) showed that the high voltage exponential decay pulses did not induce permanent modifications of ovalbumin in solution since the observed structural changes of ovalbumin were completely reversed after storage for 24 h at 4 °C. These results indicate that at high voltage, like for the WPs, the duration of the PEF treatment is an important determinant for the influence on the structure and functional properties of the egg white proteins.

2.2.1.3 Meat and Fish Proteins

Meat proteins consist of a large number of proteins that are usually clustered in three groups: (1) myofibrillar, (2) stromal, and (3) sarcoplasmic proteins having significant differences in structure and functional properties (Kang & Singh, 2015). Myofibrillar proteins are the most abundant proteins in the muscle (60%) and consist mainly of actin (13%) and myosin (26%). These two proteins form during muscle contraction a more complex protein known as actomyosin. Loss of myofibrillar integrity post-mortem has been associated with degradation of troponin-T and desmin (Suwandy, Carne, van de Ven, Bekhit, & Hopkins, 2015a). Stromal proteins (e.g. collagen, elastin and reticulin) exist in the connective tissue. Collagen is converted to gelatin during cooking, whereas elastin is not. Sarcoplasmic proteins include the hemoglobin and myoglobin pigments contributing to the red color of muscle.

The effects of PEF on meat and fish products have received some attention especially in the last years (Arroyo, Eslami, et al., 2015; Arroyo, Lascorz, et al., 2015; Bekhit, Suwandy, Carne, van de Ven, & Hopkins, 2016; Bhat, Morton, Mason, & Bekhit, 2018; Gudmundsson & Hafsteinsson, 2001; Khan et al., 2018; Ma et al., 2016; O'Dowd, Arimi, Noci, Cronin, & Lyng, 2013; Suwandy et al. 2015a, b, c, d; Töpfl, 2006). Most of these studies focused on assessing the effect of PEF on the quality characteristics of different muscles and only a few evaluated the changes on individual proteins. Table 2.3 presents a summary of the studies performed in which proteins have been studied. No study was found on isolated meat or fish proteins. Initial studies using SDS-PAGE, by Faridnia et al. (2014) and Gudmundsson and Hafsteinsson (2001), showed that PEF treatment up to 18.6 kV/cm and 7 pulses did not affect the proteins, as the same protein bands were visible in the PEF-treated and control samples. However, more recent studies aiming to understand the improvement in meat tenderness caused by PEF treatments, have reported changes in the troponin-T and desmin proteins in PEF treated beef muscles during aging (Bekhit et al., 2016; Suwandy et al., 2015a, b, c, d). Using SDS-PAGE and Western blotting to assess the impact of variable PEF conditions, aging time, and type of meat, these studies showed increased proteolysis of troponin-T and desmin (and improved tenderness) in PEF-treated beef compared with control. The changes in proteins are probably caused by permeabilization of cells and release of proteolytic enzymes.

Table 2.3 Effect of pulsed electric field treatments on meat and fish proteins

Protein	Product	Treatment intensity	Effects	Reference
Meat protein profile	Beef	0.2–0.6 kV/cm for 20 μs; T ≤ 33.20 ± 4.05 °C	No changes in protein profile	Faridnia, Bekhit, Niven, and Oey (2014)
	Salmon	1.36 kV/cm, 40 pulses for 80 μs at room temperature	No denaturation of proteins	Gudmundsson and Hafsteinsson (2001)
	Cod	10.6–18.6 kV/cm, 7 pulses (pulse time not provided)		
	Chicken	1.36 kV/cm, 40 pulses for 80 μs at room temperature		
	Lumpfish roes	12 kV/cm, 2–12 pulses (pulse time not provided)		
Troponin-T and desmin	Beef	5 kV, 10 kV × 20, 50, 90 Hz, T not provided	Increased proteolysis of troponin-T and desmin for PEF treatments at 5 kV–90 Hz, 10 kV–20 Hz and 10 kV–50 Hz samples	Suwandy et al. (2015a)
	Beef	5 kV, 10 kV × 20, 50, 90 Hz, T not provided	Significant proteolysis of troponin-T and desmin for (5–10 kV)/20 Hz	Suwandy et al. (2015b)
	Beef	10 kV, 90 Hz, 20 μs, T not provided	Higher increase in proteolysis in low-pH (5.5–5.8) than high-pH (>6.1) samples	Suwandy et al. (2015c)
	Cold-boned beef loins and topsides	Repeat PEF at (10 kV, 90 Hz, 20 μs)	Increased proteolysis of troponin T and desmin after one treatment but decreased for 2 and 3 repeated treatments	Suwandy et al. (2015d)
	Hot-boned beef loins and topsides	Repeat PEF at (10 kV, 90 Hz, 20 μs); T: 24.8 ± 1.4 °C and 24.5 ± 1.0 °C for loins and topsides, respectively	Decreased proteolysis of troponin T in repeated treatments	Bekhit et al. (2016)

2.2.1.4 Plant-Based Proteins

Very little is known about the effect of PEF on the proteins of plant-based foods, even though a few studies have been performed with soy proteins. Soybean proteins consist mainly of storage globulins, which are divided into two subgroups: glycinin and β-conglycinin. These two proteins differ significantly with respect to their functional properties, especially gelation. Gels made from glycinin are harder than gels from β-conglycinin due to differences in the structure of these proteins. Glycinin is a hexamer with a molecular weight of 300–380 kDa while β-conglycinin is a glycoprotein with a molecular weight of 150–200 kDa (Mojica, Dia, & Mejía, 2015).

Effects of PEF on the functional properties of soy protein isolate (SPI) reported by Li and Chen (2006) showed an increase in the degree of denaturation and aggregation and, consequently, a decrease in protein solubility (above 40 kV/cm or 432 μs), emulsifying capacity (above 30 kV/cm or 144 μs), foaming capacity (above 35 kV/cm or 432 μs), and surface hydrophobicity (above 30 kV/cm or 288 μs). However, Li, Chen, and Mo (2007) showed that PEF treatment (0–40 kV/cm for 0–547 μs) did not induce any significant changes to the secondary structure of soybean proteins in SPI solutions in water at pH 8. In contrast, Liu, Zeng, Deng, Yu, and Yamasaki (2011), using Fourier transform infrared spectroscopy, reported that PEF treatment above 35 kV/cm induced bond vibration within amino acid side chains, anti-parallel β-sheets, β-turn, and β-sheets, suggesting unfolding of the secondary structure of proteins in SPI solutions in water at pH 7. Furthermore, complete denaturation of β-conglycinin and glycinin in SPI were also observed in the same study by differential scanning calorimetry. The difference in results may also be due to the lab procedures used for extraction and production of SPI that may also have an effect on the subsequent changes in proteins during PEF.

To conclude, studies performed regarding the effects of PEF on food proteins are limited and few have used advanced measurement techniques to understand protein modifications at the molecular level. A number of contradictory results have been published, that may be due, on one side, to differences in equipment and processing conditions used, that may have influenced the distribution of the electric field and temperature in the product, and on the other side, to product properties such as composition, pH, concentration of proteins, pre-treatments, and structure/food matrix that influence the extension of protein modifications and interaction with other components. Studies so far, show that PEF, depending on the processing conditions (intensity of electric field, duration of treatment and product temperature), has the ability to alter the protein structure directly or through release of proteolytic enzymes ultimately affecting the functional properties and quality of the food product, but the exact mechanisms have not been elucidated.

2.2.2 Lipids and Oils

All plant- and animal-based foods contain lipids, with concentrations varying between 0.3% in most fruits and vegetables up to 65% in walnuts and from 1% in human milk to 15% in milk from the Black bear. Lipids play a major role in human nutrition, as source of energy and essential fatty acids and carrier of fat-soluble vitamins, and they have also an important role in food structure contributing to attractive sensorial attributes of foods. During processing, lipids undergo oxidation, hydrolysis, and thermal decomposition, which have a negative effect on food quality.

2.2.2.1 Milk Fat

The fat in fresh bovine whole milk exists as fat globules (MFGs) encapsulated within a native stabilizing membrane known as the milk fat globule membrane (MFGM) (Fox, Uniacke-Lowe, McSweeney, & O'Mahony, 2016). These globules are composed mainly of triglycerides (≈98%) and have a diameter between 0.1 and 15 µm depending on cow feed and other factors (Bermúdez-Aguirre, Fernández, Esquivel, Dunne, & Barbosa-Cánovas, 2011; Garcia-Amezquita, Primo-Mora, Barbosa-Cánovas, & Sepulveda, 2009; Walstra, 1983). The MFGM originates from the endoplasmatic reticulum membrane in the secretory cell, forming a small vesicle that is surrounded by the outer cell membrane during secretion (Michalski, Michel, Sainmont, & Briard, 2002). The structure of the MFGM, as proposed by Lopez et al. (2011), is shown in Fig. 2.2. The outer double layer membrane is, similarly to other biological cell membranes, primarily composed of proteins, glycoproteins, polar lipids, phospholipids, cholesterol, and other minor components (Sharma, Oey, & Everett, 2014). The MFGM provides stability to the fat globules suspended in the aqueous phase of milk by preventing flocculation of fat globules and by protecting the triglycerides from hydrolysis by indigenous or bacterial lipases (Keenan & Patton, 1993). Since PEF is known to affect biological membranes (Liu et al., 2015), this technique is expected to affect the membranes surrounding the MFGs and thus various properties of the fat.

Up to now, only a limited number of studies have been published on PEF-induced effects on milk fat. The studies show that various properties of the milk fat can be affected by the PEF treatment, as summarized in Table 2.4.

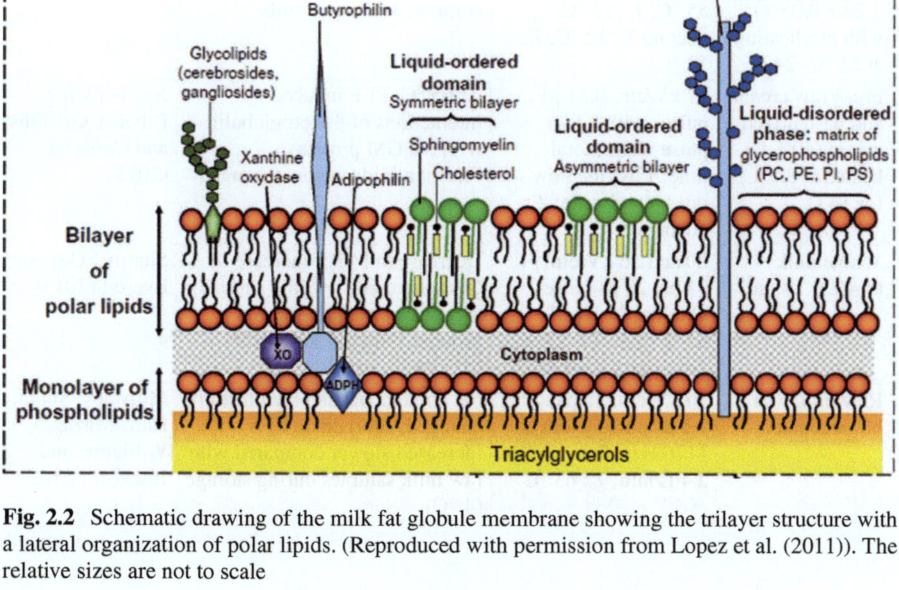

Fig. 2.2 Schematic drawing of the milk fat globule membrane showing the trilayer structure with a lateral organization of polar lipids. (Reproduced with permission from Lopez et al. (2011)). The relative sizes are not to scale

Table 2.4 Summary of reported effects of the pulsed electric field (PEF) on milk fat

Product	PEF processing conditions	Effects on milk fat	Reference
Raw milk	20–80 kV/cm; 1–10 μs; 2 μs width pulses; T_p: 55 °C	No change in fat integrity	Dunn (1996)
β-lactoglobulin model emulsions with skim, whole milk, cream	21–36 kV/cm; 200 pulses of 0.8–1.6 μs at 1 Hz; T_o: <30 °C	MFG size did not change in skimmed milk & whole milk. In cream, larger MFG dissociated	Barsotti et al. (2001)
Beverage: 50% orange juice, 20% UHT skimmed milk	35 & 40 kV/cm; 40–180 μs; Flow rate: 60 mL/min	Non-significant changes in the contents of SFA, MUFA or PUFA. A small reduction in fat content	Zulueta, Esteve, Frasquet, and Frígola (2007)
Whole milk	36 & 42 kV/cm with 24–64 & 8–24; 2.6 μs; flow rate: 383.3 mL/min; T_o: <25 °C	No significant effect on MFGs particle size distribution. MFG size changes were similar to LTLT	Garcia-Amezquita et al. (2009)
Skimmed (0.21% fat) and whole milk (3.94% fat)	30–54 kV/cm; 2 μs pulse width; pulses: 12 to 30; T_p: 20–40 °C	Increase or decrease in the fat content of skim and whole milk depending on T and intensity of electric fields	Bermúdez-Aguirre et al. (2011)
Skim milk	15–20 kV/cm; 0.33 μF; 20–60 exponentially decaying pulses at 0.50 Hz; T_o: <35 °C	The apparent viscosity of sample increased caused by an increase in the surface area of MFGs	Xiang, Simpson, Ngadi, and Simpson (2011)
Fresh, whole milk (fat content 3.85 ± 0.05 to 4.20 ± 0.05%) with pre-heating at 55 °C- 24 s	16–26 kV/cm; 85 ms pulses of 20 μs at 10–60 Hz; First T_i: 55 °C, T_o: 12 °C; Second T_i: 12 °C, T_o: 12 °C;	26 kV/cm for 34 μs at 55 °C reduced xanthine oxidase (32%) and lipolysable fat (82%) compared with raw milk	Sharma et al. (2014)
Fresh raw cream and pasteurized cream (40% fat, heated at 80 °C for 16 s)	37 kV/cm; 0.31 μF; pulses with 1.5 μs pulse width; total time: 1705 μs; flow rate: 25 mL/min; T_p: 50 and 65 °C	At 65 °C: PEF induced interactions of β-lactoglobulin with MFGM proteins. Phospholipids were unchanged	Xu, Walkling-Ribeiro, Griffiths, and Corredig (2015)
Whole milk (4.4 ± 0.1% fat)	20 and 26 kV/cm; 34 μs; 20 μs pulse width at 20 Hz; T_i: 55 °C; T_o: ~17–22 °C flow rate 4.2 mL/s	Decrease in MFG, increase in ζ-potential and specific surface area. Adsorption of plasma proteins onto the surface of MFGM occurred.	Sharma, Oey, and Everett (2015)
Raw milk was homogenized	30 kV/cm; 22 μs; 1.5 μs pulse width at 1176 Hz); flow rate: 2.4 L/min; T_o: 63 °C	Fat content was not significantly changed; short chain FFA increased slower compared with raw milk samples during storage (4 °C)	McAuley, Singh, Haro-Maza, Williams, and Buckow (2016)

(continued)

Table 2.4 (continued)

Product	PEF processing conditions	Effects on milk fat	Reference
Whole milk	20–26 kV/cm; 34 μs; (pulses 20 μs width at 20 Hz;) T_i: 55 °C; T_o: ~23.1 ± 1.6 °C flow rate of 4.2 mL/s	Whey proteins adsorbed onto the surface of MFGs; denaturation and surface hydrophobicity of proteins from MFGM surface increased with the PEF treatment intensity	Sharma, Oey, and Everett (2016)
Beverage (juices with a pasteurized whole (3.5% fat) or skim (0.3% fat) milk	35 kV/cm; 1800 μs; (pulses of 4 μs width at 200 Hz); T_p: <40 °C Flow rate of 60 mL/s	After PEF only linoleic acid increased (20%). Lower amount of palmitic, linoleic and linolenic acids (12–20%) were found at day 56	Salvia-Trujillo, Morales-de la Peña, Rojas-Graü, Welti-Chanes, and Martín-Belloso (2017)

T_i inlet temperature, T_p processing temperature, T_o outlet temperature, *LTLT* low temperature long time, *MFG* milk fat globule, *MFGM* milk fat globule membrane, *FFA* free fatty acids, *SFA* saturated fatty acids, MUFA monounsaturated fatty acids, PUFA polyunsaturated fatty acids

Varying results have been reported regarding the effect of PEF on MFG and the consequences for the determination of the fat content in milk and milk containing drinks. Zulueta et al. (2007) studied the effect of PEF processing (at 35 & 40 kV/cm, 2.5 μs) of an orange juice-milk (20% UHT-milk) beverage, and found that the fat content in the juice-milk beverage was reduced (*p* < *0.05*) after 60 min of treatment, at both electrical field strengths, and they speculated that the reduction of the size of the MFGs was negatively influencing the extraction of the fat content. Bermúdez-Aguirre et al. (2011) investigated the effect of PEF treatment with various voltages and number of pulses and also at various temperatures (20–40 °C) on the fat content in skimmed milk (0.21% fat) and whole milk (3.94% fat). After PEF treatment of whole milk, in most cases, the fat content was reduced. Likewise, for skimmed milk, PEF treatment at 31–46 kV/cm, during 2 μs, and at 20 °C also caused a decrease in the fat content. This decrease was explained by electrodeposition of milk constituents on the electrodes. On the other hand, PEF treatment of skimmed milk at 40 °C and high voltage caused a significant (*p* < *0.05*) increase in the fat content. The authors speculated that this may be due to the breakdown or electroporation of the milk fat globule membrane (MFGM) releasing triacylglycerols to the medium, which might cause a slightly higher quantification of fat (Bermúdez-Aguirre et al., 2011). At other PEF processing conditions, there was no specific trend in the fat content for the two types of milk. This is in line with a very recent study (McAuley et al., 2016) reporting that the fat content in whole milk was not significantly affected by PEF treatment (at 30 kV/cm, 22 μs), at either 53 or 63 °C. The variation in the results regarding the impact of fat content is likely due to variations in the type of PEF equipment and processing conditions, as well as in fat extraction/analysis methods.

To our knowledge, no studies have been performed on the effect of PEF treatment on the fatty acid composition in the milk; however, few studies have reported the effects of PEF on fatty acids in milk containing beverages. Zulueta et al. (2007) reported no significant changes in the contents of saturated fatty acids, monounsaturated fatty acids, or polyunsaturated fatty acids after PEF treatment of an orange juice-milk beverage. Neither peroxides nor intolerable levels of furfurals were detected. This is in line with the results obtained by Salvia-Trujillo et al. (2017) using a similar juice-milk beverage fortified with n-3 fatty acids and oleic acids. They found, however, an increase in some of the fatty acids after storage at 4 °C for 56 days. Since PEF treatment partially inactivates enzymes and microorganisms in the milk, the differences in the content of free fatty acids during storage was explained by the residual enzymatic (e.g. lipase, or microbial) activity. The partial inactivation of lipase in bovine whole milk has also been reported by Sharma et al. (2014), who found that PEF treatments (up to 26 kV/cm for 34 μs up to 55 °C for 24 s) reduced the activity of lipase detected by a reduction of lipolysable fat. Regarding microbial growth, McAuley et al. (2016) found that three volatile short chain fatty acids in homogenized and PEF-treated (30 kV/cm; 22 μs; at 63 °C) milk increased at a slower rate during refrigerated (4 °C) storage, compared to in raw milk, which was ascribed to the pasteurizing effect of PEF (i.e. inactivation of microorganisms producing the FFA).

PEF treatment of milk may also influence the size of the MFG, however, conflicting results have been reported. In some studies, PEF treatment was found to decrease the MFG size, whereas, in a few others, the size of MFGs was unaltered or increased. This may be due to different processes taking place simultaneously, e.g. dissociation of fat globules and association of fat globules with other MFG or with proteins in milk, respectively. For example, Barsotti et al. (2001) reported that PEF treatment at 29 kV/cm did not induce any changes in MFG size of whole milk, whereas PEF treatment at 32 kV/cm tended to dissociate fat globule aggregates in cream. Also, when treated at even higher voltages (36 kV/cm and 42 kV/cm, up to 64 pulses), PEF processing did not modify the true mean diameter of MFG, since the ζ-potential values were unchanged, but it induced small globules to clump together rather than coalesce, causing an apparent increment in the population of larger milk-fat globules in whole milk (Garcia-Amezquita et al., 2009). This is also supported by other studies indicating no damage to the MFGM after various PEF treatments of milk (Dunn, 1996; Sharma et al., 2016; Xu et al., 2015).

However, the association of MFGs into larger clumps may also be caused by attachment of whey proteins and casein micelles onto the surface of MFG, suggesting that some alterations in the MFGM may take place after all. For example, Xiang, Simpson, et al. (2011) reported that the apparent viscosity of reconstituted skimmed milk increased with increasing electric field intensity (15, 18 and 20 kV/cm) and the number of pulses (20, 40 and 60), which was ascribed to the intermolecular interactions as a result of the attractions between adjacent denatured milk protein and MFGs, leading to an increase in the fat globule size. The adsorption of both casein micelles and whey proteins to the MFGM in PEF-treated milk or cream has been confirmed by several studies using different methods (e.g. by confocal microscopy and DSC measurements) (Sharma et al., 2014, 2016; Xu et al., 2015). Sharma et al. (2015) also found adsorption of WP and caseins onto the surface of the MFGM during pre-heating and PEF treatments. However, in their study, PEF treatment caused an increase in the ζ-potential and specific surface area, and a decrease in the size of the MFGs, which they ascribed to the shear forces generated during pumping in their PEF equipment. Anyway, the authors considered that the PEF treatment (20 or 26 kV/cm for 34 ms, 4.2 mL/s) caused less damage to MFG surfaces than thermal treatment and suggested that these processes are slightly different (Fig. 2.3). PEF treatment may induce conformational changes in the globular proteins exposing hydrophobic sites through which adsorption onto the MFGM surface may occur through hydrophobic association, and like after thermal treatments, whey proteins may also be covalently bound onto the surface of MFGMs through disulphide linkages formed with cysteine-containing MFGM proteins (Sharma et al., 2016).

2.2.2.2 Egg and Meat Lipids

Unlike milk lipids, no studies were found regarding the effect of PEF on lipids in eggs. With respect to meat, the consequences of PEF on meat oxidation have been reported in turkey breast meat by Arroyo, Eslami, et al. (2015) and Arroyo, Lascorz, et al. (2015) using 300 pulses of 20 μs at 7.5–12.5 kV (fresh meat), 14–25 kV (frozen meat) and in lamb by Ma et al. (2016) using 1–1.4 kV/cm, 90 Hz, 20 ms. In these studies, lipid oxidation was determined by measuring the thiobarbituric acid-reactive substances and no information was provided about changes in fatty acids. No studies were found on lipids in fish treated by PEF.

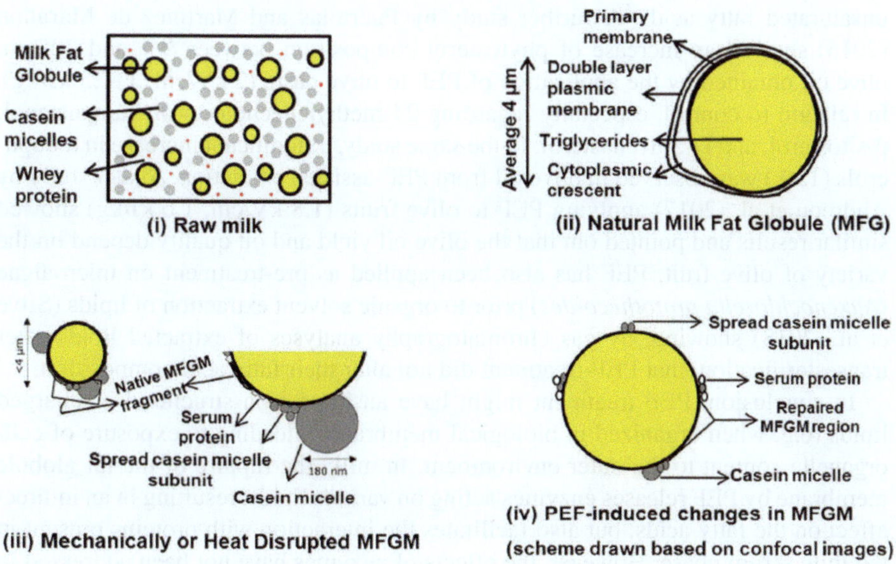

Fig. 2.3 Milk fat globules (MFGs) in: (i) raw milk, (ii) native MFG, (iii) disruption by mechanical or heat treatment; (iv) disruption due to pulsed electric fields. (Reproduced with permission from Sharma et al. (2014))

2.2.2.3 Plant-Based Oils

The application of PEF for treatment of plant-based lipids or plant oils have received very little attention. Zeng, Han, and Zi (2010) assessed the effect of different PEF treatments (20–50 kV/cm) on the physicochemical properties of peanut oil during storage up to 100 days at 40 °C. The GC/MS analysis showed no significant changes in peanut oil composition directly after PEF treatment, however, during the 100 days storage, slight differences in unsaturated and saturated fatty acids were observed between PEF-treated and untreated oils. The authors suggested that PEF treatment might reduce the rate of lipid oxidation and improve the shelf life of the oil-rich product.

Most of the studies reporting on effect of PEF treatment on lipids concern studies where this technology was used to improve the extraction of oils from maize, soybeans, or rapeseeds (Guderjan, Elez-Martínez, & Knorr, 2007; Guderjan, Töpfl, Angersbach, & Knorr, 2005), olives (Abenoza et al., 2013; Andreou et al., 2017; Puértolas & Martínez de Marañón, 2015) and algae (Silve et al., 2018). For example, Guderjan et al. (2005) showed that pulsed electric fields (0.6–1.3 kV/cm, 20–50 pulses) applied to hulled and dehulled rapeseed increased the yield and the concentrations of total tocopherols and phytosterols in the oil. No effects on unsaturated lipids and saponification values were determined, but more free fatty acids were obtained after pressing of the PEF-treated dehulled rapeseed. Abenoza et al. (2013) studying the impact of PEF treatments on olive paste (0–2 kV/cm) at different malaxation times (0, 15, and 30 min) and temperatures (15 and 26 °C) showed no differences in fatty acid composition or content of saturated, unsaturated and polyunsaturated fatty acids. A further study by Puértolas and Martínez de Marañón (2015) showed an increase of phytosterol composition between 9% and 20% in olive oil obtained by the application of PEF to olive paste (2 kV/cm; 11.25 kJ/kg), in relation to control, especially regarding 24-methylenecholesterol, campestanol, β-sitosterol, and D-5-avenasterol. In the same study, a significant increase in tocopherols (15%) was observed in olive oil from PEF-assisted extraction. A later study by Andreou et al. (2017) applying PEF to olive fruits (1.8 kV/cm, 1.6 kJ/kg) showed similar results and pointed out that the olive oil yield and oil quality depend on the variety of olive fruit. PEF has also been applied as pre-treatment on microalgae (*Auxenochlorella protothecoides*) prior to organic solvent extraction of lipids (Silve et al., 2018) showing, by gas chromatography analyses of extracted lipids after transesterification, that PEF-treatment did not alter their fatty acid composition.

In conclusion, PEF treatment might have an impact on structured or charged lipids (e.g. when organized in biological membranes), leading to exposure of cell/organelle content to the outer environment. In milk, the rupture of the fat globule membrane by PEF releases enzymes acting on various lipids, resulting in an indirect effect on the fatty acids, but also facilitates the interaction with proteins present in the milk serum phase. However, the effects of enzymes have not been addressed in most of the studies. The use of PEF as pre-treatment of oilseeds has shown no negative effects on the quality and functionality of oils at field intensities required for permeabilization of cellular structures.

2.2.3 Carbohydrates

PEF treatment may not directly affect carbohydrates since most of them are non-charged. Accordingly, only a few studies have focused on the effect of PEF on properties of carbohydrates. Regarding disaccharides, Garde-Cerdán, Arias-Gil, Marsellés-Fontanet, Ancín-Azpilicueta, and Martín-Belloso (2007) found no significant reduction in sugar content of grape juice after applying PEF (12 kV /cm for 20 μs).

Nevertheless, recent studies with polysaccharides have shown that PEF treatment has an effect on starch, causing damage to the starch granules and their crystalline structure. Studies have been performed with corn (Han, Zeng, Zhang, & Yu, 2009), potato (Han, Zeng, Yu, Zhang, & Chen, 2009), tapioca (Han et al., 2012), maize (Han et al., 2012), and rice (Zeng, Gao, Han, Zeng, & Yu, 2016) starch in water suspensions showing some differences depending on the type of starch. Han, Zeng, Zhang, et al. (2009) showed that PEF treatment (up to 50 kV/cm) caused dissociation, denaturation, and damage of starch granules, which lead to a decrease of gelatinization temperature and enthalpy with the increase of electric field strength. Further studies with maize starch using thermogravimetric analysis (TGA) and Nuclear Magnetic Resonance (NMR) analysis showed no significant difference between native and PEF-treated starch indicating that PEF treatments may not affect the chemical structure of maize starch (Han, Yu, et al., 2012). More recently, Zeng et al. (2016) using small angle X-ray scattering observed that the relative crystallinity of PEF-treated rice starches (up to 50 kV/cm) decreased with an increase of electric field intensity, thus increasing the amount of rapidly digestible starch and decreasing the slowly digestible starch. Reduction of molecular weight of amylopectin has also been observed in tapioca (Han, Zeng, et al., 2012) and rice starch (Zeng et al., 2016). BeMiller (2018) discussed the physical processes for modification of starch and concluded that the limited information available indicates that the damage of starch granules and their crystalline structure by PEF makes the starch more susceptible to enzymatic hydrolysis.

In conclusion, further studies are needed to clarify the mechanisms of modification of starch by PEF. The permeabilization of amyloplast membrane by PEF or heat-induced effects on proteins naturally associated with starch granules are likely to influence the overall properties of starch.

2.2.4 Bioactive Compounds

Foods not only supply macronutrients such as lipids, proteins, and carbohydrates; they are also important sources of vitamins, antioxidants and other compounds (e.g. secondary metabolites and peptides) with an important health impact. Vitamins, especially water-soluble vitamins are sensitive to heat treatment (Rechcigl, 1984), but likely to be better preserved by PEF. When PEF applied to foods is not accompanied by extensive heat treatment, vitamins and other health-promoting compounds are not expected to be affected to a great extent by the treatment.

2.2.4.1 Milk Components

Bovine milk contains all the nutrients and vitamins needed for the growth of the calf. Milk and dairy products contribute a large part of the human intake of riboflavin (vitamin B_2), cobalamin (vitamin B_{12}) and vitamin A, but also contain other vitamins such as water-soluble thiamine (vitamin B_1), ascorbic acid (vitamin C), fat-soluble cholecalciferol (vitamin D), and tocopherol (vitamin E). Actually, there is a lot of provitamin A, which contributes to the yellow color of these products.

Despite the potential advantages of PEF in preserving vitamins, studies are scarce. Bendicho, Espachs, Arántegui, and Martín (2002) reported that no changes were observed in either water-soluble (riboflavin, thiamine) or fat-soluble vitamins (cholecalciferol and tocopherol) in milk after applying PEF treatments (22.6 kV/cm for 400 μs). However, a slight decrease in ascorbic acid was observed. Riener, Noci, Cronin, Morgan, and Lyng (2009) did not find any changes in thiamine, riboflavin, and retinol in skimmed milk. More studies are needed regarding the effect of PEF processing conditions in preserving the vitamins in animal-based products.

2.2.4.2 Plant Compounds

Apart from vitamins, plants contain glycosides and polyphenolic compounds which act as antioxidants and may help to improve health. When producing juice from plants (e.g. tomatoes or berries), PEF can help extracting such compounds to the juice. Table 2.5. presents a summary of studies on the effect of PEF treatments on ascorbic acid, phenolic compounds, as well as on other compounds such as carotenoids and chlorophyll.

Studies published so far have shown that PEF-treated products have higher retention of vitamins than heat-treated products. For example, Cortés et al. (2006) reported a decrease of 7.5% in vitamin A in PEF-treated (30 kV/cm for 100 μs) orange juice, which is lower than the loss of 15.6% after heat treatment (90 °C for 20 s). Torregrosa, Cortés, Esteve, and Frígola (2005) also found that PEF-treated orange-carrot juice (25–30 kV/cm 30–340 μs) had a higher vitamin A concentration than that found in the heat pasteurized juice.

The contents of ascorbic acid in orange and grape juices are almost not affected by PEF treatment when combining the treatments at 80 kV/cm with bacteriocin (Hodgins et al., 2002; Wu et al., 2005). Several research groups have found that changes in ascorbic acid content of heat-pasteurized liquid plant-based products (orange juice, strawberry juice, apple cider and "gazpacho" soup) are between 2 and 3 times higher than those of PEF-processed products (Elez-Martínez & Martín-Belloso, 2007; Elez-Martínez, Soliva-Fortuny, & Martín-Belloso, 2005; Evrendilek et al., 2000; Odriozola-Serrano et al., 2008a, 2008b; Qiu, Sharma, Tuhela, Jia, & Zhang, 1998). Elez-Martínez and Martín-Belloso (2007) reported that pulses applied in a bipolar mode, as well as lowering field strength, treatment time, pulse frequency and width, lead to higher levels of ascorbic acid retention in both orange juice and "gazpacho" soup. Vallverdú-Queralt et al. (2012) and Odriozola-Serrano

Table 2.5 Effect of pulsed electric field treatments on bioactive compounds in plant-based systems

Compound	Product	Treatment intensity	Effects	Reference
Ascorbic acid	Tomato fruit	5–30 pulses at 0.4–2.0 kV/cm	98.3 % retention	Vallverdú-Queralt et al. (2012)
	Tomato juice	35 kV/cm for 1500 μs; T ≤40 °C	86.5 % retention	Odriozola-Serrano et al. (2008b)
	Orange juice	20 pulses at 80 kV/cm; T = 42–44 °C	97.5 % retention	Hodgins et al. (2002)
		15–35 kV/cm for 100–1000 μs; T ≤35 °C	87.5–98.2 % retention	Elez-Martínez and Martín-Belloso (2007)
	Strawberry juice	35 kV/cm for 1700 μs; T ≤40 °C	98 % retention	Odriozola-Serrano et al. (2008a)
	Grape juice	20 pulses at 65–80 kV/cm T increase negligible	No change	Wu et al. (2005)
	Apple juice and cider	22–35 kV/cm for 94–166 μs; T = 26–27 °C	No change	Evrendilek et al. (2000)
	'Gazpacho' soup	15–35 kV/cm for 100–1000 μs; T ≤35 °C	84.3–97.1 % retention	Elez-Martinez and Martin-Belloso (2007)
Carotenoids	Orange juice	25–40 kV/cm for 30–340 μs; T = 37–72 °C	No significant changes in overall content. Better stability of individual compounds compared to thermal treatment	Cortés et al. (2006)
	Tomato juice	35 kV/cm for 1500 μs; T ≤20 °C	Higher overall content. Better stability of individual compounds compared to thermal treatment	Odriozola-Serrano et al. (2009)
Lycopene	Tomato fruit	5–30 pulses at 0.4–2.0 kV/cm	7–45% increase in content	Vallverdú-Queralt et al. (2012)
	Tomato juice	35 kV/cm for 1500 μs; T ≤40 °C	110% retention	Odriozola-Serrano et al. (2008b)
		40 kV/cm for 57 μs; T = 45 °C, T_o = 25 °C	No significant difference	Min et al. (2003)
Flavonoids	Orange juice	35 kV/cm for 750 μs; T ≤50 °C	No changes in either individual flavanones nor in total content	Sánchez-Moreno et al. (2005)

(continued)

Table 2.5 (continued)

Compound	Product	Treatment intensity	Effects	Reference
Anthocyanins	Cranberry juice	20–40 kV/cm for 50–150 μs; T = 15–25 °C in Jin & Zhang (1999); T = 26–27 °C in Evrendilek et al. (2001).	no noticeable changes	Jin & Zhang (1999) and Evrendilek et al. (2001)
	Strawberry juice	35 kV/cm for 1700 μs; T ≤40 °C	96.5 % retention	Odriozola-Serrano et al. (2008a)
Ellagic acid	Strawberry juice	35 kV/cm for 1700 μs; T ≤40 °C	97.8 % retention	Odriozola-Serrano et al. (2008a)
Polyphenol	Tomato fruit	5–30 pulses at 0.4–2.0 kV/cm, No T information	0.6–32 % increase in content	Vallverdú-Queralt et al. (2012)
Chlorophyll	Spinach puree	20–100 kV/cm for 50–150 μs, No T information	Increased stability by increasing electric fields adding zinc ion and stabilizers	Yin et al. (2007)

et al. (2008a) also showed that high retention of ascorbic acid is found for PEF-treated tomato fruit or juice.

Vallverdú-Queralt et al. (2012) found that a 44.6% increase in total polyphenol content of tomato fruit was achieved under moderate intensity pulsed electric fields (1 kV/cm, 4 μs). However, studies on individual phenolic compounds did not show a significant effect. For example, Sánchez-Moreno et al. (2005) reported that neither changes in the total flavanones were observed, nor in the individual flavanone glycosides and their aglycons hesperetin and naringenin in orange juice by PEF treatment at 35 kV/cm for 750 μs. Likewise, PEF treatment of cranberry juice or strawberry juice did not cause any noticeable changes in anthocyanins and ellagic acid although they are particularly sensitive to heat treatments and oxidations (Evrendilek et al., 2001; Jin & Zhang, 1999; Odriozola-Serrano et al., 2008b).

Dietary chlorophyll derivatives in spinach puree, which have been identified as potential chemopreventative agents with antioxidant and antimutagenic activities, have been treated by a combination of PEF treatments (20–100 kV/cm) and the addition of water-soluble Zn^{2+} concentrations (20–200 ppm). Electric field strengths above 60 kV/cm showed to be detrimental to the color of spinach puree, but a satisfactory color in the juice was obtained with zinc concentrations below 75 ppm without significantly affecting flavor (Soliva-Fortuny, Balasa, Knorr, & Martín-Belloso, 2009; Yin et al., 2007). The mechanisms underlying these changes are however not fully understood.

In conclusion, as shown in the examples mentioned above, the use of PEF for pasteurization has the advantage over heat treatment to preserve vitamins and other health-promoting compounds. Most studies have been performed in juices, and studies in other products are needed. The use of PEF at low intensity to enhance the production of bioactive compounds in plants by stress response is limited but could be a relevant technology to enhance the nutritional value in foods. Further studies are still needed to elucidate the mechanisms underlying the preservation or enhancement of bioactive components, specifically in more complex food matrices where interaction with the other components is likely to occur.

2.2.5 Flavor Compounds

2.2.5.1 Milk

The effects of PEF treatment on volatile compounds are usually dependent on the type of compound and the limited information available does not provide an understanding of the changes reported. Zhang et al. (2011), using GC-MS, observed that PEF-treated milk (30 kV/cm) had a higher content of pentanal, hexanal, and nonanal, while methyl ketones, heptanal, and decanal contents were lower than in heat pasteurized milk. On the other hand, PEF-treated milk showed no significant differences from pasteurized milk with respect to contents of acids (e.g. acetic acid, butanoic acid, hexanoic acid, octanoic acid, and decanoic acid), lactones and alcohols, whereas 2(5H)-furanone was only detected in the PEF-treated milk. Although GC-MS detected differences between pasteurized and PEF-treated milk, GC-olfactometry showed no significant difference between the two milk samples. It cannot be excluded that the differences in microbial inactivation between PEF and heat pasteurized milk may be the reason for the changes in flavor compounds observed.

2.2.5.2 Plant-Based Foods

In plant-based product, flavor has great importance for the consumers' acceptability, especially for juice. Several studies have compared the PEF-treated juice with thermally treated juice with respect to loss of flavor components. However, little is known about the effects of the intensity of PEF treatment on the release of flavor compounds from the food matrix, and the mechanism and factors that take part in the release of these compounds (Soliva-Fortuny et al., 2009).

For example, the flavor of orange juice consists of more than 200 aroma compounds of different chemical nature. The contents of some hydrophobic compounds (e.g. limonene, myrcene, valencene, and α-pinene) involved in the flavor were found to remain similar or even higher (18–32%) after PEF processing (35 kV/cm for 59 μs), due to a release of these components from their hydrophobic environment by the treatment (Ayhan, Zhang, & Min, 2002). More polar compounds such as octanal, decanal, linalool, and ethyl butyrate remain unchanged or with just a little loss (Ayhan et al., 2002; Jin & Zhang, 1999).

In general, a loss of 3–9% is found in juices after PEF-treatment whereas a reduction of 22% of the volatile compounds has been reported for heat pasteurization (Soliva-Fortuny et al., 2009). The losses from PEF treatment could probably be minimized through better control of treatment temperature, which occasionally reaches values close to 60 °C in continuous-flow pilot plant equipment (Mañas & Vercet, 2006).

Regarding orange juice, Min, Jin, Min, Yeom, and Zhang (2003) and Yeom, Streaker, Zhang, and Min (2000) reported a higher content of flavor components (α-pinene, myrcene, octanal, limonene, and decanal) in PEF-treated orange juice (35–40 kV/cm for 59–97 μs) than in heat-pasteurized juice (94.6 °C for 30 s). Consistently, the flavor components of other citrus juices (e.g. orange, lemon, grapefruit, and tangerine) are not lost to a significant degree due to PEF treatments (Cserhalmi, Sass-Kiss, Tóth-Markus, & Lechner, 2006).

Similarly to what has been described for orange juice, PEF-treated tomato juice retain the characteristic flavor-related compounds (e.g. *trans*-2-hexenal, 2-isobutylthiazole, *cis*-3-hexanol) compared with thermal processing (Min, Jin, Min, et al., 2003). Likewise, for cranberry juice, which is characterized by its special flavor, PEF-treated juices could not be distinguished from untreated juices in terms of retention of volatile compounds, indicating that PEF treatment did not alter the flavor or aroma profile of cranberry juice (Jin & Zhang, 1999).

In conclusion, studies so far have shown that PEF technology preserves the flavor compounds in liquids (juices) to a higher extent than conventionally applied heat treatments. Although, kinetic data at different PEF processing conditions is still limited and few studies have focused on changes during storage. The formation or degradation of some flavor compounds observed in different juices are now well understood and may be associated with the microbial or enzymatic activity.

2.3 Food Microstructure

A number of studies have reported on permeabilization and disintegration of cell membranes in plant and animal tissues by the application of PEF treatment (Angersbach, Heinz, & Knorr, 2000; Fincan & Dejmek, 2002; Fincan, DeVito, & Dejmek, 2004; Gudmundsson & Hafsteinsson, 2001; Knorr & Angersbach, 1998; Lebovka, Bazhal, & Vorobiev, 2001; Lebovka, Praporscic, & Vorobiev, 2004). When the transmembrane potential conferred by PEF on the cell exceeds the critical value, rupture of the cell membrane occurs, and the inner contents of the cell are released, producing an osmotic imbalance between the internal and external surroundings of the cell which leads to swelling and eventually death. Depending on the intensity of the electric field applied, cell electroporation or rupture can be either reversible or irreversible (Vorobiev & Lebovka, 2009). Living cells have repair mechanisms, which enable them to recover after PEF treatment, but these mechanisms are still not fully understood. Levine and Vernier (2010) performed molecular dynamic simulations to explain the formation and annihilation of pores in a mixture of water and phospholipid bilayers by electric fields, but no study has been done in biological membranes.

2.3.1 Animal-Based Foods

A number of studies have demonstrated physical changes in meat structure caused by PEF (Arroyo, Lascorz, et al., 2015; Bekhit et al., 2016; Faridnia et al., 2014; Gudmundsson & Hafsteinsson, 2001). Application of PEF to meat permeabilizes the cells of the muscle fibers, which as consequence improves or in turn improves the efficiency of meat processing, as well as the quality and the functional attributes, such as tenderness, color, and water-holding capacity. As described in previous sections, the disruption of cell structures by PEF accelerates meat maturation, water removal during drying and marinating/curing processes. However, the intensity of the PEF treatment needed to accomplish the acceleration of such processes should not cause negative effects on the microstructure of foods that are strongly associated with sensorial attributes of foods.

For example, extensive microscopy characterization performed by Gudmundsson and Hafsteinsson (2001) showed that PEF treatment (1.36 kV/cm and 40 pulses each pulse is 2 μs) caused a size reduction of 39% in chicken muscle cell but without causing a visible change in the appearance. Compared with chicken, salmon was more affected by PEF treatment. Treatment of salmon with 1.36 kV/cm and 40 pulses caused leakage of collagen from the extracellular space. Cryo-SEM images of PEF-treated beef (*M. longissimus thoracis*) (0.3–0.6 kV/cm, 8.5–34 kJ/kg) showed that the meat became more porous as the electric field strength increased (Faridnia et al., 2014). According to Töpfl (2006), PEF treatment can reduce the drying time of salted pork shoulder by 80% (from 300 to 60 h) dependent on treatment intensity and salting procedure. Töpfl and Heinz (2007) showed that a PEF treatment (3 kV/cm, 5 kJ/kg) prior to immersion in brine (0.08 kg/kg nitrite salt) could improve the diffusion of salt and nitrate in pork haunches.

2.3.2 Plant-Based Foods

Permeabilization of the cell membranes by application of PEF to cellular plant-based foods, like fruits and vegetables, causes softening of the tissue and, as previously discussed in this chapter, it improves the extraction ability of bioactive compounds and oils and enhances mass transfer (Shynkaryk et al., 2009). As shown in Fig. 2.4, Töpfl (2006) reported that the cell disintegration of apple tissue is highly related to the intensity of the electric field and the number of pulses. The energy required for cell permeabilization is dependent on the type of plant material, and it has been reported to be 6.4–16.2 kJ/kg for potatoes (Angesbach & Knorr, 1997), 0.4–6.7 kJ/kg for grape skins (López, Puértolas, Condón, Álvarez, & Raso, 2008), 2.5 kJ/kg for red beetroot and 3.9–7 kJ/kg for sugar beet (López, Puértolas, Condón, Raso, & Alvarez, 2009; López, Puértolas, Condón, Raso, & Álvarez, 2009).

The softening of plant-based foods caused by PEF treatments have been industrially explored in cutting and/or slicing operations since it significantly reduced the

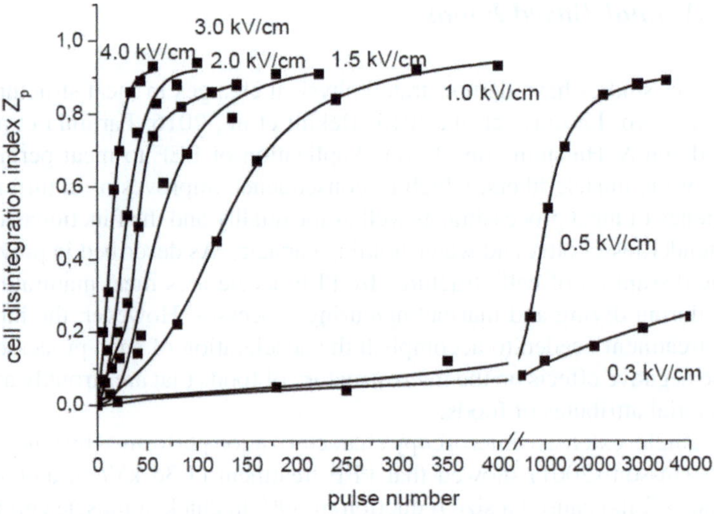

Fig. 2.4 Permeabilization of apple cells as function of electric field intensity. (With permission of Töpfl (2006))

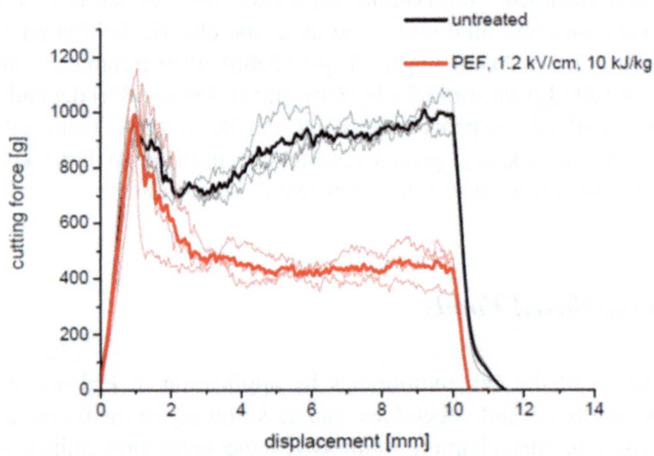

Fig. 2.5 Effects of PEF on cutting force of potato tissue compared with untreated tissue. (Reproduced with permission from Töpfl (2006))

energy required and extended operation time of knives used for cutting. Töpfl (2006) reported a German study performed by Kraus showing an improvement of cut quality and reduction of approx. 50% cutting force in sugar beet. Further studies by Töpfl (2006) reported similar results for potatoes (Fig. 2.5). In addition, to a reduction in cutting force, PEF pre-treatment before slicing produced smooth and flat cutting surfaces, which led to a decrease in oil uptake in frying (Janositz, Noack, & Knorr, 2011).

Although PEF has obvious effects on cell permeabilization, there are few studies dedicated to understanding their consequences on the microstructure of plants tissues. Faridnia, Burritt, Bremer, and Oey (2015) performed a comprehensive study to gain an in-depth understanding of the effect of PEF (0.2–1.1 kV/cm, 50 Hz, 1–10 kJ/kg) on the microstructure of potato tubers. Cell viability, leakage of ions, and microstructure damages were assessed using tetrazolium salt staining, atomic absorption spectrophotometry, field emission scanning electron microscopy and energy scattering spectrometer (FESEM-EDS) analysis, and Cryo-SEM, respectively. The results showed that PEF caused an uneven distribution of cell damage due to the differences in cellular structures within the potato tuber. Wiktor et al. (2016) showed that small cavities, a compact structure, and a high density characterized the microstructure of dried carrots not treated by PEF, while carrots treated with PEF (0–5 kV/cm, pulses 0–100, 0–80 kJ/kg) had larger cavities, especially when a higher intensity of the electric field was applied.

In conclusion, PEF has a well-documented effect on the permeabilization of membranes that surround animal and plant cells and their organelles. The permeabilization of membranes causes in efflux of metabolites from the cell contents/organelles to the outer environment, modifying the ionic strength/pH, facilitating many enzymatic reactions and promoting interactions between components that in natural structures are separated. These changes may have a positive, negative or neutral effect on the final quality of the product, depending on the specific product and further processing. Reported studies have determined the critical electric field intensity for a number of food matrices.

In plant foods, permeabilization of the cell membrane causes loss of turgor pressure leading to softening of plant tissues. This effect is the base for successful commercial application of PEF for softening of plant tissues (e.g. before potatoes slicing for production of chips). In general, limited fundamental studies have been performed to understand structural changes at the microstructural level and their consequences for the release of enzymes or substrates that may enhance quality degradation reactions.

2.4 Sensorial Attributes and Consumer Acceptance

A major advantage of pulsed electric fields over conventional heat treatment is the possibility to extend the shelf life of food or increase extraction yield without the development of undesirable changes in nutritional and sensorial quality as caused by conventional heat treatments. Sensorial attributes of foods, such as color, odor, texture, and flavor are important for consumer acceptance and successful industrial implementation of PEF.

Despite initial concerns regarding the safety of PEF-treated products, since 2002, and especially in the last 5 years, a significant number of studies have reported on sensorial evaluation by human panels of PEF-treated products and overall better acceptance of the products in comparison with thermal-treated ones. Table 2.6 presents a

Table 2.6 Effect of pulsed electric field treatments on sensorial attributes of foods

Product	PEF processing conditions	Sensorial Test	Impact on product	Reference
Juices and Beverages				
Apple juice	Combination of PEF (24 kV/cm or 34 kV/cm for 89 μs) and high-intensity light pulses (pulse length 360 μs, frequency 3 Hz) and exposed to energy dosages of 5.1 J/cm^2 or 4.0 J/cm^2)	9-point hedonic scale using 31 untrained panelists (20 males and 11 females) to evaluate the color, odor, sweetness, acidity and overall acceptability	Variation between the sensorial attributes tested depending on processing conditions used	Caminiti et al. (2011)
Apple juice	24.8 kV/cm, 60 pulses, 169 ms treatment time, 53.8 °C	Triangle difference and ranking preference tests. 25 untrained Judges.	A significant difference in overall flavor between PEF- and thermally-processed juices, with a preference for the PEF	Sulaiman, Farid, and Silva (2017)
Apple juice	15.5 kV/cm, 48 Hz 20 μs,16 L/h, specific energy 158 kJ/l, T:30 °C	Consumers' sensory perceptions by napping with ultra-flash profiling; Four trials, 8–10 participants/trial (17 m and 21 fem; Chinese ethnicity), Not trained panel	PEF-processed juices had fresh flavor in comparison with high pressure and thermally treated juices	Lee, Kebede, Lusk, Mirosa, and Oey (2017)
Apple cider	23 kV/cm. T: 48 °C	Triangle test detect differences in aroma	No significant difference between PEF-treated and or heat-treated apple cider. PEF cider has greater retention of apple aroma volatiles.	Azhu Valappil, Fan, Zhang, and Rouseff (2009)
Apple cider	5, 10, 13, 17, 21 and 23 kV/cm for 150 μs; 48 °C	Triangle test with 50 untrained panelists	PEF-treated cider was preferred over thermal and UV cider at the end of the storage period.	Azhuvalappil, Fan, Geveke, and Zhang (2010)
Melon and watermelon juices	35 kV/cm for 1709 μs at 193 Hz and 4 μs pulse duration) and watermelon (35 kV/cm for 1682 μs at 193 Hz and 4 μs pulse duration) juices. Addition of citric acid or cinnamon bark oils as antimicrobials	Hedonic scale - 30 untrained panelists evaluated preference of odor, color, taste, sourness and overall acceptability	Negative effects on taste, odor, and sourness attribute when citric acid or cinnamon bark oil is added to the juices at concentrations ensuring its safety	Mosqueda-Melgar, Raybaudi-Massilia, and Martín-Belloso (2008)

Orange juice	35 kV/cm for 59 μs	9- hedonic scale test 12- member untrained panel.	Loss of octanal and decanal compounds had no significant effect on the sensorial quality of orange juice	Ayhan et al. (2002)
Orange Juice	40 kV/cm for 97 ms	9-point hedonic test with 30 untrained panelists rate the preference of color, appearance, texture, flavor, and overall acceptability.	Texture, flavor, and overall acceptability were ranked highest for fresh juice, followed by PEF-processed juice and then by thermally processed juice	Min, Jin, Min, et al. (2003)
Orange Juice	Sonication (TS) at 55 °C for 10 min followed by continuous PEF at a field strength of 40 kV/cm for 150 s.	9-point hedonic scale test with 37 (16 female, 21 male) untrained panelists ranging from 18 to 65 years of age. For color, odor, flavor and overall acceptability	All sensory attributes were rated equivalent for TS/PEF- and HTST-treated juice ($P \geq 0.05$)	Walkling-Ribeiro, Noci, Cronin, Lyng, and Morgan (2010)
Orange juice	13–25 kV/cm treatment time 1034 to 1206 us) T: 31 to 42 °C	10- point hedonic scale with 13 untrained panelists evaluated (turbidity, color, odor, aroma and general acceptance) were used to test orange juice samples	The PEF-treated samples had higher sensory scores than the thermally-treated juices	Agcam, Akyıldız, and Akdemir Evrendilek (2014)
Pomegranate juice	0, 17, 23, and 30 kV/cm 108.4 μs and T 5, 15, 25, and 35 °C (37.5, 50.3, and 65.3 J)	9-point hedonic scale with 30 trained panelists, assessing for flavor, color, taste, sourness, aftertaste, and overall acceptance.	Sensorial attributes were not significantly affected by PEF alone nor PEF combined with mild heat treatment	Evrendilek (2017)
Pomegranate Juice	35 and 38 kV/cm for 281 μs at 55 °C with a flow rate of 100 L/h	9-point hedonic scale test with 30 consumers. Pomegranate juice evaluated by overall appearance, flavor, acceptability	No statistical differences between PEF-treated and fresh juices. Consumers preferred unpasteurized and PEF compared with thermally processed samples.	Guo et al. (2014)

(continued)

Table 2.6 (continued)

Product	PEF processing conditions	Sensorial Test	Impact on product	Reference
Tomato Juice	40 kV/cm, for 57 μs. T 45 °C	9-point hedonic test with twenty-eight of 30 trained panelists, rate the preference of appearance, color, texture, flavor, and overall acceptability	PEF processed juices were preferred over cold break juice	Min, Jin, and Zhang (2003)
Tomato Juice	40 kV/cm for 57 μs	A 9-point hedonic scale and a triangle test with 30 untrained panelists triangle test	The flavor of PEF-processed juice was preferred to hot break juice	Min and Zhang (2003)
Watermelon juice	11 kV/cm; 175 kJ/kg; T 35 °C	Triangle tests (DIN standard EN ISO 4120). A non-trained sensory with 30 panelists, observe, smell and taste the juices	Significant differentiation between unprocessed and processed juices. No preference test.	Aganovic et al. (2016)
Blended orange and carrot juice	Treatment 1, 25 kV/cm and 280 ms, Treatment 2, 25 kV/cm and 330 ms,	25 untrained volunteers. Odor and taste compared with the untreated juice control	Sensory characteristics of the PEF-treated juice similar to fresh and HTST-pasteurized juice	Rivas, Rodrigo, Martínez, Barbosa-Cánovas, and Rodrigo (2006)
Apple and cranberry juice blend	34 kV/cm, 18 Hz, 93 μs combined with UV and High-Intensity Light Pulses (HIPL) and Manothermosonication (MTS)	9-point hedonic scale test with 35 untrained panelists, evaluating appearance, odor, flavor, sweetness, acidity and overall acceptability	UV + PEF and HILP + PEF did not impact odor and flavor, while combinations with MTS adversely affected these attributes	Caminiti et al. (2011)
Citrus juices (orange, grapefruit, tangerine)	28 kV/cm; 100 μs (50 pulses 2 μs) 84 mL/min, T: 20–27 °C	Sensory properties evaluated by electronic tongue and electronic-nose	Electronic nose and – tongue able to differentiate PEF treated samples from fresh juice	Hartyáni et al. (2011)
Sour cherry juice apricot and peach nectars	24 kV/cm; total time 66–210 μs 50-mL/min flow rate, T = 23 °C	9-point hedonic scale for flavor, color, taste, consistency, sourness, sweetness, aftertaste and overall acceptance. 60 untrained panelists	The PEF treatment preserved 94% of the sensory properties. Increasing treatment time did not significantly affect any of the sensory scores	Evrendilek, Avsar, and Evrendilek (2016)

Peach nectar	Different treatment times (0, 66, 131 and 210 μs) 50 mL/min	9-point hedonic scale with 30 trained panelists for flavor, color, taste, consistency, sourness, sweetness, aftertaste and overall acceptance	PEF treatment significantly enhanced or did not affect (88%) the sensory properties for treatments of >131 μs	Evrendilek et al. (2016)
Dairy pudding (yogurt, fruit jelly and corn syrup)	30 kV/cm electric field strength for 32 μs total treatment	9-point hedonic test with an untrained 30 to 59 panelists evaluated appearance, texture, flavor, taste, and overall acceptability	No significant difference between the control and processed products	Yeom, Evrendilek, Jin, and Zhang (2004)
Yogurt-based drink (yogurt, water, sugar strawberry)	60 °C for 30 s and 30 kV/cm, 100 L/h	9-point hedonic scale test with an untrained sensory panel with 20 students, evaluation of appearance, texture, flavor, taste, and overall acceptability,	The control and 60°C + PEF samples revealed no significant difference in the selected sensory attributes	Evrendilek, Yeom, Jin, and Zhang (2004)
Smoothie beverage	Thermal (55 °C after 60 s) and PEF (34 kV/cm, 60 μs)	9-point hedonic scale by an untrained panel of 35 (19 female, 16 male) 18 and 65 years	Mild pasteurization and combination of heat and PEF resulted in smoothies with comparable overall acceptability	Walkling-Ribeiro et al. (2010)
Wine	33 kV/cm and an energy output of 158 ± 2.8 kJ/kg, a treatment time of 105 ± 2.8 μs and a pulse width of 8 μs. Ti =18 °C Tmax = 48 °C	10- expert judges with a wide wine tasting experience, UNE-EN ISO: 8586–2: 2009. 6 female and 4 men, 27–60 years. Descriptive and triangle test (standard UNE-EN ISO 4120:2008.	For three out of the four treated wines, the sensorial quality was better after the PEF treatment	González-Arenzana et al. (2018)
Wine	17, 24 and 31 kV/cm at 30 °C	on a 9-point hedonic test by 50 trained panelists involving visual evaluation, smell, and taste,	No significant difference in sensory attributes between control and PEF-treated wine	Abca and Evrendilek (2015)

(continued)

Table 2.6 (continued)

Product	PEF processing conditions	Sensorial Test	Impact on product	Reference
Beer (0 and 7% alc/vol beers and 4 and 7% alc/vol beers)	T < 43 °C, 45 kV/cm electrical field intensity, 46 pulses, 70 μs)	Aroma blindly evaluated by 5 trained panelists who smelled beer to detect the lightstruck smell. Flavor of each beer on a 5-point scale	PEF beers did not develop the lightstruck character, being acceptable in terms of sensory attributes	Milani, Alkhafaji, and Silva (2015)
Beer	10.5 mL/s flow rate, 22 kV/cm electric; 154 s total treatment time. Temperature from 4 to 9 °C	9-point hedonic scale test with 25 trained sensory panelists for flavor/aroma, mouthfeel, foam condition, color, and overall acceptance	Significantly lower rating for flavor and mouth feeling of the PEF-treated beer	Evrendilek, Li, Dantzer, and Zhang (2004)
Applesauce	60 kV; 82 and 109 μs (200 J/mL) combined with a high-temperature short time (70–80 °C) for approx. 70 s	9-point Hedonic scale test with 34–41 untrained panelists consumers. Evaluation of appearance, color, odor, overall flavor, natural fruit flavor, texture, and overall quality	Sensory quality of PEF applesauce was affected by storage time and temperature and depended on the apple variety.	Jin et al. (2009)
Meat and Eggs				
Beef muscle	1.4 kV/cm; 600 pulses of 20 μs; 10 Hz)	9-point hedonic scale. The untrained consumer panel consisted of 20 untrained people balanced for age and gender	PEF treatment of beef has no impact on consumer's sensory acceptance. No differences on odor 60% scored PEF treated samples as tender versus 27.5% for untreated samples	Arroyo, Lascorz, et al. (2015)
Turkey breast meat	4.4–12 kV (20 μs), frequency (5 Hz), of 26, 78, 114, and 194 kJ/kg.	Consumer acceptability for color, texture, and odor. 40 untrained panelists	Slight differences in texture and odor between the PEF treated and controls	Arroyo, Eslami, et al. (2015)
Lamb meats	1–1.4 kV/cm, the specific energy of 88–109 kJ/kg	Ten trained panelists (5 males, 5 females, 21–29 years of age; Sensorial attributes tested (initial tenderness, meaty, browned, juicy, livery, oxidized, warmed-over flavor)	Tenderness of cooked meat was not significantly affected by PEF treatment, but PEF affected the temporal flavor profiles of meaty and oxidized flavor attributes	Ma et al. (2016)

Liquid whole egg	25 kV/cm and 200 kJ/kg) followed by heat (60 °C during 3.5 s) + GRAS-considered compounds mandarin EO, lemon EO or (+)-limonene	9-point hedonic scale for omelets and sponge cakes made with PEF treated eggs; Untrained panel with 20 (omelets) and 65 (sponge cakes) volunteers	Sensory acceptable omelets and sponge cakes with PEF treated eggs. No effect of antimicrobial compounds.	Espina, Monfort, Álvarez, García-Gonzalo, and Pagán (2014)
Pre-treatment in the production of apple juice or cider				
Apple mash (apple juice)	3 kV/cm and the specific energy input was 10 kJ/kg	Triangle test with 30 trained panelists	Sensory properties of the juices did not significantly differ when PEF pre-treatment was applied	Schilling et al. (2008)
Apple mash (Apple cider and juice)	Treatment of the mash (32 ms and 46 kJ/kg)	9 points hedonic test with 12 trained panelists (AFNOR norm V09–105), description of color, odor, taste, and aroma	The color of the PEF treated juices was greatly appreciated by the judging panel	Turk, Billaud, Vorobiev, and Baron (2012)
Apple mash	650 V/cm, 23.2 ms, 32 kJ/kg)	9-point hedonic scale 12 trained panelists (4 men, 8 women, and aged 46–74 years), evaluated color, odor, taste, and aroma	PEF treatment had a positive impact on the sensory attributes. Higher intensity of typical odor of apples	Turk, Vorobiev, and Baron (Turk, Vorobiev, & Baron, 2012
Pre-treatment of grapes before wine production				
Grapes	5 kV/cm/3.67 kJ/kg at a frequency of 122 Hz. T increase <2 °C	Wine evaluated after 8 months in a bottle using a triangle test by a trained panel in two sessions	No sensory differences in color and bouquet were detected after 8 months of aging in bottle	Puértolas, Saldaña, Álvarez, and Raso (2010)
Grape Pomace	5 kV/cm (total specific energy: 3.67 kJ/kg) at a frequency of 122 Hz	A panel of 18 trained panel (7 m 11 f); ages (21–45 y). Color intensity, astringency, bouquet, flavor, and global evaluation. 5 points scale	No significant differences in sensory attributes between wines were detected	Puértolas, Hernández-Orte, Sladaña, Álvarez, and Raso (2010)
Grapes	4.3 kV/cm, 60 μs) at a flow of 1900 kg/h	Triangle tests with 9 trained panelists	Wine from PEF-treated grapes was significantly preferable to the other two wines	Luengo, Franco, Ballesteros, Álvarez, and Raso (2014)

(continued)

Table 2.6 (continued)

Product	PEF processing conditions	Sensorial Test	Impact on product	Reference
Grape mash	7 kV/cm; Pulses of 150 μs	A panel made up of 15 judges consisting of students	The PEF red wine showed higher scores for all sensory attributes compared with control	Vicaş et al. (2017)
Pre-treatment of olive past before olive oil production				
Olive paste (Arbequina)	0–2 kV/cm at different malaxation times (0, 15, and 30 min) and temperatures (15 and 26 °C)	Sensory analysis by EEC/2568/91 and EEC/640/2008. Olive oil evaluated by 10 trained panelists according to positive and negative attributes	PEF treatment did not generate any bad flavor or taste in the olive oil	Abenoza et al. (2013)
Olive paste	2 kV/cm; 11.25 kJ/kg	Sensory analysis Regulation (EC) 2568/1991. 12 trained panelists smelled and tasted the oil samples, using a 10-cm scale	PEF had no negative effects on sensory characteristics of the olive oil, maintaining the highest quality according to EU legal standards (EVOO; extra virgin olive oil)	Puértolas and Martínez de Marañón (2015)

summary of the studies performed. Most commonly, a 9-point hedonic scale test has been used with untrained assessors, but recently triangle tests and use of trained panels have also been used. Juices, nectars, and beverages are by far the type of food products most evaluated by sensorial panels. Tests performed with apple juice and apple cider showed enhanced flavor of PEF-treated juices compared with heat-treated juice (Azhu Valappil et al., 2009; Lee et al., 2017; Sulaiman et al., 2017), but the results depend on the processing conditions used in the PEF treatment (Azhu Valappil et al., 2009; Caminiti, Noci, et al., 2011; Caminiti, Palgan, et al., 2011).

Regarding orange juice, the sensorial advantages of PEF compared to thermal treatment are less evident. Although Agcam et al. (2014), Ayhan et al. (2002), and Min, Jin, Min, et al. (2003) reported that orange juice had higher sensorial scores than pasteurized juice, the differences were relatively small. Compared with fresh juice, Walkling-Ribeiro, Noci, Cronin, Lyng, and Morgan (2009) observed no significant differences from juices treated with PEF combined with sonication. Furthermore, as reported by Buckow, Ng, and Toepfl (2013), orange juice thermally treated at 98 °C for 11 s did not show significant flavor changes compared to fresh orange juice, therefore tangible advantages may be largely dependent on the quality of the fresh juice. Two studies with pomegranate juice showed no statistical differences between PEF-treated and fresh juices, and consumers preferred fresh and PEF treated to the thermally processed ones (Evrendilek, 2017; Guo et al., 2014).

The same trend is observed for the other juices, i.e. depending on the processing conditions used, the sensory characteristics of the PEF-treated juices are similar to fresh juices and better or equivalent to pasteurized ones. Although it is likely that PEF has clear advantages in keeping the aroma and flavor compounds of juices, it is important to note that, in many studies, the microbial inactivation of thermally-treated juices and PEF-treated ones is not equivalent and therefore, the tangible advantages cannot be fully assessed.

Moreover, recent studies on the application of PEF to alcoholic beverages, like wine (Abca & Evrendilek, 2015; González-Arenzana et al., 2018) and beer (Milani et al., 2015) have shown the potential of this technology in controlling the microbial activity without affecting the sensorial aspects.

The application of PEF to meat products is more complex due to the different types of meat products, and the limited number of studies available does not allow extrapolating conclusions. Arroyo, Eslami, et al. (2015) observed that PEF treatment of beef has no impact on consumer's sensory acceptance (no differences on odor) and 60% of panelists scored PEF-treated samples as tender in comparison with 27.5% for untreated samples. Further studies, by the same authors (Arroyo, Lascorz, et al., 2015) with turkey breast meat, and by Ma et al. (2016) with lamb, showed only slight differences in texture and odor between the PEF-treated and untreated, except for an increase in oxidation taste in the PEF-treated samples. However, treatment of eggs by PEF (in combination with the addition of antimicrobials) produced omelets and sponge cakes with acceptable sensorial quality (Espina et al., 2014).

The remaining studies reported in the literature were focused on sensorial evaluation of products where PEF has been used as a pre-treatment before the production of apple juice or cider (Schilling et al., 2008; Turk, Billaud, et al., 2012; Turk, Vorobiev, & Baron, 2012), wine (Luengo et al., 2014; Puértolas, Hernández-Orte, et al., 2010; Puértolas, Saldaña, et al., 2010; Vicaş et al., 2017), or olive oil (Abenoza et al., 2013; Puértolas & Martínez de Marañón, 2015). In all these studies, no significant differences or improvement of the sensory attributes were observed due to the use of PEF. Worth noting is that the application of PEF to grapes before wine production seems to be a promising technology to improve the quality of wines in terms of color, taste and high phenolic content in oak-aged red wines (Luengo et al., 2014; Puértolas, Saldaña, et al., 2010; Vicaş et al., 2017).

In conclusion, sensorial evaluation of PEF-treated foods by trained panels is essential to assess the acceptability of the PEF technology. A number of studies have been performed showing a very good potential of this technology for pasteurization of juices and extraction of oils, wine, and some juices.

2.5 Conclusions and Outlook

Understanding the effects of pulsed electric fields at the molecular level is still an emerging research area. Relatively few studies have focused on the effects of PEF on proteins, lipids or polysaccharides with the purpose to explore molecular or microstructural modifications of these macromolecules. Contradictory results are often reported in the literature, due to differences in PEF equipment used, electric fields applied and probably due to more or less associated thermal effects. A better understanding of electric field distributions in foods and differentiation between thermal and electric fields effects is necessary to get mechanistic insights into the effect of PEF on macromolecules. It is necessary also the use advanced *in situ* methods for characterization of dynamic changes in these macromolecules, and the reversibility of such reactions.

Reported studies have demonstrated the impact of PEF on permeabilization of cell membranes and the critical electric field intensity has been determined for a number of food matrices. Thus, there is evidence about the effects of PEF on disturbing phospholipid bilayers causing the formation and further development of a pore, that can be reversible or not. Modification of the fat globule membrane in milk is another example where PEF may influence the adsorption of proteins on the surface of the fat globules, leading to changes in the functionality of dairy products.

Softening of plant tissues before slicing and application of PEF as pre-treatment to enhance extraction before the production of wine and olive oil are two successful examples where permeabilization of plant cell membranes provides tangible advantages in terms of product quality and process efficiency.

Modification of proteins by PEF is much less understood. Reported studies have shown modifications in the secondary structure that are strongly dependent on the intensity and duration of electric field strength. However, the temperature increases during PEF treatment or the characteristics of the protein solution or protein ingredients such as conductivity, pH, concentration or thermal stability, is often not considered. More fundamental studies are still needed using the advanced methodology for protein characterization to understand how PEF can initiate modification in proteins and the consequences for protein functionality. In complex food matrices, the effect of proteolytic enzymes needs to be considered, as they will contribute to such protein modifications.

Modification of polysaccharides (i.e. starch) by PEF has been the focus of a number of recent studies that have shown that dissociation, denaturation, and damage of starch granules lead to a decrease of gelatinization temperature and enthalpy. Full understanding of these changes has not been achieved.

Regarding minor compounds, such as flavor compounds, vitamins, and other bioactive compounds, studies so far showed that for some products, PEF provides clear advantages in comparison to thermal processing. However, with exception of juices, very limited studies are available to provide the required evidence.

Sensorial attributes of PEF-treated foods evaluated by trained panels has provided new insight on the consumer acceptance and tangible advantages of PEF compared with alternative technologies, and further studies are needed for successful commercialization of this technology.

In conclusion, PEF has the potential for modification of macromolecules and microstructures creating a new base for development of ingredients or food products with tailored functionalities. Further research is needed to elucidate the mechanism of PEF-induced modifications in macromolecules, food microstructures and consequence for functional properties. Studies with a model or single molecule system, as well as with complex formulations or real foods are needed to understand the relationships between macromolecules and structure formation.

References

Abca, E. E., & Evrendilek, G. A. (2015). Processing of red wine by pulsed electric fields with respect to quality parameters. *Journal of Food Processing and Preservation, 39*(6), 758–767.

Abenoza, M., Benito, M., Saldaña, G., Álvarez, I., Raso, J., & Sánchez-Gimeno, A. C. (2013). Effects of pulsed electric field on yield extraction and quality of olive oil. *Food and Bioprocess Technology, 6*(6), 1367–1373.

Aganovic, K., Grauwet, T., Siemer, C., Toepfl, S., Heinz, V., Hendrickx, M., & Loey, A. V. (2016). Headspace fingerprinting and sensory evaluation to discriminate between traditional and alternative pasteurization of watermelon juice. *European Food Research and Technology, 242*(5), 787–803.

Agcam, E., Akyıldız, A., & Akdemir Evrendilek, G. (2014). Comparison of phenolic compounds of orange juice processed by pulsed electric fields (PEF) and conventional thermal pasteurisation. *Food Chemistry, 143*, 354–361.

Andreou, V., Dimopoulos, G., Alexandrakis, Z., Katsaros, G., Oikonomou, D., Toepfl, S., … Taoukis, P. (2017). Shelf-life evaluation of virgin olive oil extracted from olives subjected to nonthermal pretreatments for yield increase. *Innovative Food Science & Emerging Technologies, 40*, 52–57.

Angersbach, A., Heinz, V., & Knorr, D. (2000). Effects of pulsed electric fields on cell membranes in real food systems. *Innovative Food Science & Emerging Technologies, 1*(2), 135–149.

Angesbach, A., & Knorr, D. (1997). Anwendung elektrischer Hochspannungsimpulse als Vorbehandlungsverfahren zur Beeinflussung der Trocknungscharakteristika und Rehydratationseigenschaften von Kartoffelwürfeln. *Food/Nahrung, 41*(1), 194–200.

Arroyo, C., Eslami, S., Brunton, N. P., Arimi, J. M., Noci, F., & Lyng, J. G. (2015). An assessment of the impact of pulsed electric fields processing factors on oxidation, color, texture, and sensory attributes of turkey breast meat. *Poultry Science, 94*(5), 1088–1095.

Arroyo, C., Lascorz, D., O'Dowd, L., Noci, F., Arimi, J., & Lyng, J. G. (2015). Effect of pulsed electric field treatments at various stages during conditioning on quality attributes of beef *longissimus thoracis* et *lumborum* muscle. *Meat Science, 99*, 52–59.

Ayhan, Z., Zhang, Q. H., & Min, D. B. (2002). Effects of pulsed electric field processing and storage on the quality and stability of single-strength orange juice. *Journal of Food Protection, 65*(10), 1623–1627.

Azhu Valappil, Z., Fan, X., Zhang, H. Q., & Rouseff, R. L. (2009). Impact of thermal and nonthermal processing technologies on unfermented apple cider aroma volatiles. *Journal of Agricultural and Food Chemistry, 57*(3), 924–929.

Azhuvalappil, Z., Fan, X., Geveke, D. J., & Zhang, H. Q. (2010). Thermal and nonthermal processing of apple cider: Storage quality under equivalent process conditions. *Journal of Food Quality, 33*(5), 612–631.

Barsotti, L., Dumay, E., Mu, T. H., Fernandez Diaz, M. D., & Cheftel, J. C. (2001). Effects of high voltage electric pulses on protein-based food constituents and structures. *Trends in Food Science & Technology, 12*(3), 136–144.

Bekhit, A. E.-D. A., Suwandy, V., Carne, A., van de Ven, R., & Hopkins, D. L. (2016). Effect of repeated pulsed electric field treatment on the quality of hot-boned beef loins and topsides. *Meat Science, 111*, 139–146.

BeMiller, J. N. (2018). Physical modification of starch. In M. Sjöö & L. Nilsson (Eds.), *Starch in food: Structure, function and applications*. Woodhead Publishing, an imprint of Elsevier, Duxford.

Bendicho, S., Espachs, A., Arántegui, J., & Martín, O. (2002). Effect of high intensity pulsed electric fields and heat treatments on vitamins of milk. *Journal of Dairy Research, 69*(01), 113–123.

Bermúdez-Aguirre, D., Fernández, S., Esquivel, H., Dunne, P. C., & Barbosa-Cánovas, G. V. (2011). Milk processed by pulsed electric fields: Evaluation of microbial quality, physicochemical characteristics, and selected nutrients at different storage conditions. *Journal of Food Science, 76*(5), S289–S299.

Bhat, Z. F., Morton, J. D., Mason, S. L., & Bekhit, A. E.-D. A. (2018). Current and future prospects for the use of pulsed electric field in the meat industry. *Critical Reviews in Food Science and Nutrition*. https://doi.org/10.1080/10408398.2018.1425825

Buckow, R., Ng, S., & Toepfl, S. (2013). Pulsed electric field processing of orange juice: A review on microbial, enzymatic, nutritional, and sensory quality and stability. *Comprehensive Reviews in Food Science and Food Safety, 12*(5), 455–467.

Caminiti, I. M., Noci, F., Muñoz, A., Whyte, P., Morgan, D. J., Cronin, D. A., & Lyng, J. G. (2011). Impact of selected combinations of non-thermal processing technologies on the quality of an apple and cranberry juice blend. *Food Chemistry, 124*(4), 1387–1392.

Caminiti, I. M., Palgan, I., Noci, F., Muñoz, A., Whyte, P., Cronin, D. A., … Lyng, J. G. (2011). The effect of pulsed electric fields (PEF) in combination with high intensity light pulses (HILP) on *Escherichia coli* inactivation and quality attributes in apple juice. *Innovative Food Science & Emerging Technologies, 12*(2), 118–123.

Cortés, C., Esteve, M. J., Rodrigo, D., Torregrosa, F., & Frígola, A. (2006). Changes of colour and carotenoids contents during high intensity pulsed electric field treatment in orange juices. *Food and Chemical Toxicology, 44*(11), 1932–1939.

Cserhalmi, Z., Sass-Kiss, Á., Tóth-Markus, M., & Lechner, N. (2006). Study of pulsed electric field treated citrus juices. *Innovative Food Science & Emerging Technologies, 7*(1–2), 49–54.

Dalgleish, D. G. (2011). On the structural models of bovine casein micelles - review and possible improvements. *Soft Matter, 7*(6), 2265–2272.

Dunn, J. (1996). Pulsed light and pulsed electric field for foods and eggs. *Poultry Science, 75*(9), 1133–1136.

Elez-Martínez, P., & Martín-Belloso, O. (2007). Effects of high intensity pulsed electric field processing conditions on vitamin C and antioxidant capacity of orange juice and gazpacho, a cold vegetable soup. *Food Chemistry, 102*(1), 201–209.

Elez-Martínez, P., Soliva-Fortuny, R. C., & Martín-Belloso, O. (2005). Comparative study on shelf life of orange juice processed by high intensity pulsed electric fields or heat treatment. *European Food Research and Technology, 222*(3–4), 321–329.

Espina, L., Monfort, S., Álvarez, I., García-Gonzalo, D., & Pagán, R. (2014). Combination of pulsed electric fields, mild heat and essential oils as an alternative to the ultrapasteurization of liquid whole egg. *International Journal of Food Microbiology, 189*, 119–125.

Evrendilek, G. A. (2017). Impacts of pulsed electric field and heat treatment on quality and sensory properties and microbial inactivation of pomegranate juice. *Food Science and Technology International, 23*(8), 668–680.

Evrendilek, G. A., Avsar, Y. K., & Evrendilek, F. (2016). Modelling stochastic variability and uncertainty in aroma active compounds of PEF-treated peach nectar as a function of physical and sensory properties, and treatment time. *Food Chemistry, 190*, 634–642.

Evrendilek, G. A., Dantzer, W. R., Streaker, C. B., Ratanatriwong, P., & Zhang, Q. H. (2001). Shelf-life evaluations of liquid foods treated by pilot plant pulsed electric field system. *Journal of Food Processing and Preservation, 25*(4), 283–297.

Evrendilek, G. A., Jin, Z. T., Ruhlman, K. T., Qiu, X., Zhang, Q. H., & Richter, E. R. (2000). Microbial safety and shelf-life of apple juice and cider processed by bench and pilot scale PEF systems. *Innovative Food Science & Emerging Technologies, 1*(1), 77–86.

Evrendilek, G. A., Li, S., Dantzer, W. R., & Zhang, Q. H. (2004). Pulsed electric field processing of beer: Microbial, sensory, and quality analyses. *Journal of Food Science, 69*(8), M228–M232.

Evrendilek, G. A., Yeom, H. W., Jin, Z. T., & Zhang, Q. H. (2004). Safety and quality evaluation of a yogurt-based drink processed by a pilot plant PEF system. *Journal of Food Process Engineering, 27*(3), 197–212.

Faridnia, F., Bekhit, A. E.-D. A., Niven, B., & Oey, I. (2014). Impact of pulsed electric fields and post-mortem vacuum ageing on beef *longissimus thoracis* muscles. *International Journal of Food Science & Technology, 49*(11), 2339–2347.

Faridnia, F., Burritt, D. J., Bremer, P. J., & Oey, I. (2015). Innovative approach to determine the effect of pulsed electric fields on the microstructure of whole potato tubers: Use of cell viability, microscopic images and ionic leakage measurements. *Food Research International, 77*, 556–564.

Fernandez-Diaz, M. D., Barsotti, L., Dumay, E., & Cheftel, J. C. (2000). Effects of pulsed electric fields on ovalbumin solutions and dialyzed egg white. *Journal of Agricultural and Food Chemistry, 48*(6), 2332–2339.

Fincan, M., & Dejmek, P. (2002). In situ visualization of the effect of a pulsed electric field on plant tissue. *Journal of Food Engineering, 55*(3), 223–230.

Fincan, M., DeVito, F., & Dejmek, P. (2004). Pulsed electric field treatment for solid–liquid extraction of red beetroot pigment. *Journal of Food Engineering, 64*(3), 381–388.

Floury, J., Grosset, N., Leconte, N., Pasco, M., Madec, M.-N., & Jeantet, R. (2006). Continuous raw skim milk processing by pulsed electric field at non-lethal temperature: Effect on microbial inactivation and functional properties. *Le Lait, 86*(1), 43–57.

Fox, P. F., Uniacke-Lowe, T., McSweeney, P. L. H., & O'Mahony, J. A. (2nd Eds.). (2016). *Dairy chemistry and biochemistry*. London: Springer International Publishing, Switzerland.

Garcia-Amezquita, L. E., Primo-Mora, A. R., Barbosa-Cánovas, G. V., & Sepulveda, D. R. (2009). Effect of nonthermal technologies on the native size distribution of fat globules in bovine cheese-making milk. *Innovative Food Science & Emerging Technologies, 10*(4), 491–494.

Garde-Cerdán, T., Arias-Gil, M., Marsellés-Fontanet, A. R., Ancín-Azpilicueta, C., & Martín-Belloso, O. (2007). Effects of thermal and non-thermal processing treatments on fatty acids and free amino acids of grape juice. *Food Control, 18*(5), 473–479.

González-Arenzana, L., López-Alfaro, I., Garde-Cerdán, T., Portu, J., López, R., & Santamaría, P. (2018). Microbial inactivation and MLF performances of Tempranillo Rioja wines treated with PEF after alcoholic fermentation. *International Journal of Food Microbiology, 269*, 19–26.

Guderjan, M., Elez-Martínez, P., & Knorr, D. (2007). Application of pulsed electric fields at oil yield and content of functional food ingredients at the production of rapeseed oil. *Innovative Food Science & Emerging Technologies, 8*(1), 55–62.

Guderjan, M., Töpfl, S., Angersbach, A., & Knorr, D. (2005). Impact of pulsed electric field treatment on the recovery and quality of plant oils. *Journal of Food Engineering, 67*(3), 281–287.

Gudmundsson, M., & Hafsteinsson, H. (2001). Effect of electric field pulses on microstructure of muscle foods and roes. *Trends in Food Science & Technology, 12*(3), 122–128.

Guo, M., Jin, T. Z., Geveke, D. J., Fan, X., Sites, J. E., & Wang, L. (2014). Evaluation of microbial stability, bioactive compounds, physicochemical properties, and consumer acceptance of pomegranate juice processed in a commercial scale pulsed electric field system. *Food and Bioprocess Technology, 7*(7), 2112–2120.

Han, Z., Yu, Q., Zeng, X. A., Luo, D. H., Yu, S. J., Zhang, B. S., & Chen, X. D. (2012). Studies on the microstructure and thermal properties of pulsed electric fields (PEF)-treated maize starch. *International Journal of Food Engineering, 8*(1). https://doi.org/10.1515/1556-3758.2375

Han, Z., Zeng, X., Zhang, B., & Yu, S. (2009). Effects of pulsed electric fields (PEF) treatment on the properties of corn starch. *Journal of Food Engineering, 93*(3), 318–323.

Han, Z., Zeng, X. A., Fu, N., Yu, S. J., Chen, X. D., & Kennedy, J. F. (2012). Effects of pulsed electric field treatments on some properties of tapioca starch. *Carbohydrate Polymers, 89*(4), 1012–1017.

Han, Z., Zeng, X. A., Yu, S. J., Zhang, B. S., & Chen, X. D. (2009). Effects of pulsed electric fields (PEF) treatment on physicochemical properties of potato starch. *Innovative Food Science & Emerging Technologies, 10*(4), 481–485.

Hartyáni, P., Dalmadi, I., Cserhalmi, Z., Kántor, D.-B., Tóth-Markus, M., & Sass-Kiss, Á. (2011). Physical–chemical and sensory properties of pulsed electric field and high hydrostatic pressure treated citrus juices. *Innovative Food Science & Emerging Technologies, 12*(3), 255–260.

Hemar, Y., Augustin, M. A., Cheng, L. J., Sanguansri, P., Swiergon, P., & Wan, J. (2011). The effect of pulsed electric field processing on particle size and viscosity of milk and milk concentrates. *Milchwissenschaft, 66*(2), 126–128.

Hodgins, A. M., Mittal, G. S., & Griffiths, M. W. (2002). Pasteurization of fresh orange juice using low-energy pulsed electrical field. *Journal of Food Science, 67*(6), 2294–2299.

Horne, D. S. (1998). Casein interactions: Casting light on the black boxes, the structure of dairy products. *International Dairy Journal, 8*(3), 171–177.

Janositz, A., Noack, A. K., & Knorr, D. (2011). Pulsed electric fields and their impact on the diffusion characteristics of potato slices. *LWT - Food Science and Technology, 44*(9), 1939–1945.

Jeantet, R., Baron, F., Nau, F., Roignant, M., & Brulé, G. (1999). High intensity pulsed electric fields applied to egg white: Effect on *Salmonella* Enteritidis inactivation and protein denaturation. *Journal of Food Protection, 62*(12), 1381–1386.

Jin, Z. T., Zhang, H. Q., Li, S. Q., Kim, M., Dunne, C. P., Yang, T., … Venter-Gains, J. (2009). Quality of applesauces processed by pulsed electric fields and HTST pasteurisation. *International Journal of Food Science & Technology, 44*(4), 829–839.

Jin, Z. T., & Zhang, Q. H. (1999). Pulsed electric field inactivation of microorganisms and preservation of quality of cranberry juice. *Journal of Food Processing and Preservation, 23*(6), 481–497.

Kang, I., & Singh, P. (2015). Muscle proteins. In Z. Ustunol (Ed.), *Applied food protein chemistry* (pp. 361–392). Chichester, UK: Wiley-Blackwell.

Keenan, T. W., & Patton, S. (1993). The milk lipid globule membrane. In R. G. Jensen & M. P. Thompson (Eds.), *Handbook of milk composition*. New York, NY: Academic.

Khan, A. A., Randhawa, M. A., Carne, A., Mohamed Ahmed, I. A., Al-Juhaimi, F. Y., Barr, D., … Bekhit, A. E.-D. A. (2018). Effect of low and high pulsed electric field processing on macro and micro minerals in beef and chicken. *Innovative Food Science & Emerging Technologies, 45*, 273–279.

Knorr, D., & Angersbach, A. (1998). Impact of high-intensity electric field pulses on plant membrane permeabilization. *Trends in Food Science & Technology, 9*(5), 185–191.

Lebovka, N. I., Bazhal, M. I., & Vorobiev, E. (2001). Pulsed electric field breakage of cellular tissues: Visualisation of percolative properties. *Innovative Food Science & Emerging Technologies, 2*(2), 113–125.

Lebovka, N. I., Praporscic, I., & Vorobiev, E. (2004). Combined treatment of apples by pulsed electric fields and by heating at moderate temperature. *Journal of Food Engineering, 65*(2), 211–217.

Lee, P. Y., Kebede, B. T., Lusk, K., Mirosa, M., & Oey, I. (2017). Investigating consumers' perception of apple juice as affected by novel and conventional processing technologies. *International Journal of Food Science & Technology, 52*(12), 2564–2571.

Levine, Z. A., & Vernier, P. T. (2010). Life cycle of an electropore: Field-dependent and field-independent steps in pore creation and annihilation. *The Journal of Membrane Biology, 236*(1), 27–36.

Li, S.-Q., Bomser, J. A., & Zhang, Q. H. (2005). Effects of pulsed electric fields and heat treatment on stability and secondary structure of bovine immunoglobulin G. *Journal of Agricultural and Food Chemistry, 53*(3), 663–670.

Li, S.-Q., Zhang, Q. H., Lee, Y.-Z., & Pham, T.-V. (2003). Effects of pulsed electric fields and thermal processing on the stability of bovine immunoglobulin G (IgG) in enriched soymilk. *Journal of Food Science, 68*(4), 1201–1207.

Li, Y., & Chen, Z. (2006). Effect of high intensity pulsed electric field on the functional properties of protein isolated from soybean. *Nongye Gongcheng Xuebao/Transactions of the Chinese Society of. Agricultural Engineering, 22*(8), 194–198.

Li, Y., Chen, Z., & Mo, H. (2007). Effects of pulsed electric fields on physicochemical properties of soybean protein isolates. *LWT - Food Science and Technology, 40*(7), 1167–1175.

Liu, Y. Y., Zeng, X. A., Deng, Z., Yu, S. J., & Yamasaki, S. (2011). Effect of pulsed electric field on the secondary structure and thermal properties of soy protein isolate. *European Food Research and Technology, 233*(5), 841.

Liu, Z., Hemar, Y., Tan, S., Sanguansri, P., Niere, J., Buckow, R., & Augustin, M. A. (2015). Pulsed electric field treatment of reconstituted skim milks at alkaline pH or with added EDTA. *Journal of Food Engineering, 144*, 112–118.

Lopez, C., Briard-Bion, V., Ménard, O., Beaucher, E., Rousseau, F., Fauquant, J., … Robert, B. (2011). Fat globules selected from whole milk according to their size: Different compositions and structure of the biomembrane, revealing sphingomyelin-rich domains. *Food Chemistry, 125*(2), 355–368.

López, N., Puértolas, E., Condón, S., Álvarez, I., & Raso, J. (2008). Application of pulsed electric fields for improving the maceration process during vinification of red wine: Influence of grape variety. *European Food Research and Technology, 227*(4), 1099–1107.

López, N., Puértolas, E., Condón, S., Raso, J., & Alvarez, I. (2009). Enhancement of the extraction of betanine from red beetroot by pulsed electric fields. *Journal of Food Engineering, 90*(1), 60–66.

López, N., Puértolas, E., Condón, S., Raso, J., & Álvarez, I. (2009). Enhancement of the solid-liquid extraction of sucrose from sugar beet (*Beta vulgaris*) by pulsed electric fields. *LWT - Food Science and Technology, 42*(10), 1674–1680.

Luengo, E., Franco, E., Ballesteros, F., Álvarez, I., & Raso, J. (2014). Winery trial on application of pulsed electric fields for improving vinification of garnacha grapes. *Food and Bioprocess Technology, 7*(5), 1457–1464.

Ma, Q., Hamid, N., Oey, I., Kantono, K., Faridnia, F., Yoo, M., & Farouk, M. (2016). Effect of chilled and freezing pre-treatments prior to pulsed electric field processing on volatile profile and sensory attributes of cooked lamb meats. *Innovative Food Science & Emerging Technologies, 37*, 359–374.

Mañas, P., & Vercet, A. (2006). Effect of PEF on enzymes and food constituents. In *Pulsed electric fields technology for the food industry* (pp. 131–151). Boston, MA: Springer.

Marco-Molés, R., Rojas-Graü, M. A., Hernando, I., Pérez-Munuera, I., Soliva-Fortuny, R., & Martín-Belloso, O. (2011). Physical and structural changes in liquid whole egg treated with high-intensity pulsed electric fields. *Journal of Food Science, 76*(2), C257–C264.

McAuley, C. M., Singh, T. K., Haro-Maza, J. F., Williams, R., & Buckow, R. (2016). Microbiological and physicochemical stability of raw, pasteurised or pulsed electric field-treated milk. *Innovative Food Science & Emerging Technologies, 38*, 365–373.

McMahon, D. J., & Oommen, B. S. (2013). Casein micelle structure, functions, and interactions. In P. L. H. McSweeney & P. F. Fox (Eds.), *Advanced dairy chemistry* (Vol. 1A: Proteins: Basic aspects, 4th ed., pp. 185–209). Boston, MA: Springer.

Michalac, S., Alvarez, V., Ji, T., & Zhang, Q. H. (2003). Inactivation of selected microorganisms and properties of pulsed electric field processed milk. *Journal of Food Processing and Preservation, 27*(2), 137–151.

Michalski, M.-C., Michel, F., Sainmont, D., & Briard, V. (2002). Apparent ζ-potential as a tool to assess mechanical damages to the milk fat globule membrane. *Colloids and Surfaces B: Biointerfaces, 23*(1), 23–30.

Milani, E. A., Alkhafaji, S., & Silva, F. V. M. (2015). Pulsed Electric Field continuous pasteurization of different types of beers. *Food Control, 50*, 223–229.

Min, S., Jin, Z. T., Min, S. K., Yeom, H., & Zhang, Q. H. (2003). Commercial-scale pulsed electric field processing of orange juice. *Journal of Food Science, 68*(4), 1265–1271.

Min, S., Jin, Z. T., & Zhang, Q. H. (2003). Commercial scale pulsed electric field processing of tomato juice. *Journal of Agricultural and Food Chemistry, 51*(11), 3338–3344.

Min, S., & Zhang, Q. H. (2003). Effects of commercial-scale pulsed electric field processing on flavor and color of tomato juice. *Journal of Food Science, 68*(5), 1600–1606.

Mine, Y. (2015). Egg proteins. In Z. Ustunol (Ed.), *Applied food protein chemistry* (pp. 459–490). Chichester, UK: Wiley-Blackwell.

Mojica, L., Dia, V. P., & de Mejía, E. G. (2015). Soy proteins. In Z. Ustunol (Ed.), *Applied food protein chemistry* (pp. 139–191). Chichester, UK: Wiley-Blackwell.

Mosqueda-Melgar, J., Raybaudi-Massilia, R. M., & Martín-Belloso, O. (2008). Combination of high-intensity pulsed electric fields with natural antimicrobials to inactivate pathogenic microorganisms and extend the shelf-life of melon and watermelon juices. *Food Microbiology, 25*(3), 479–491.

O'Dowd, L. P., Arimi, J. M., Noci, F., Cronin, D. A., & Lyng, J. G. (2013). An assessment of the effect of pulsed electrical fields on tenderness and selected quality attributes of post rigour beef muscle. *Meat Science, 93*(2), 303–309.

Odriozola-Serrano, I., Bendicho-Porta, S., & Martín-Belloso, O. (2006). Comparative study on shelf life of whole milk processed by high-intensity pulsed electric field or heat treatment. *Journal of Dairy Science, 89*(3), 905–911.

Odriozola-Serrano, I., Soliva-Fortuny, R., Hernández-Jover, T., & Martín-Belloso, O. (2009). Carotenoid and phenolic profile of tomato juices processed by high intensity pulsed electric fields compared with conventional thermal treatments. *Food Chemistry, 112*(1), 258–266.

Odriozola-Serrano, I., Soliva-Fortuny, R., & Martín-Belloso, O. (2008a). Changes of health-related compounds throughout cold storage of tomato juice stabilized by thermal or high intensity pulsed electric field treatments. *Innovative Food Science & Emerging Technologies, 9*(3), 272–279.

Odriozola-Serrano, I., Soliva-Fortuny, R., & Martín-Belloso, O. (2008b). Phenolic acids, flavonoids, vitamin C and antioxidant capacity of strawberry juices processed by high-intensity pulsed electric fields or heat treatments. *European Food Research and Technology, 228*(2), 239.

Perez, O. E., & Pilosof, A. M. R. (2004). Pulsed electric fields effects on the molecular structure and gelation of β-lactoglobulin concentrate and egg white. *Food Research International, 37*(1), 102–110.

Puértolas, E., Hernández-Orte, P., Sladaña, G., Álvarez, I., & Raso, J. (2010). Improvement of winemaking process using pulsed electric fields at pilot-plant scale. Evolution of chromatic parameters and phenolic content of Cabernet Sauvignon red wines. *Food Research International, 43*(3), 761–766.

Puértolas, E., & Martínez de Marañón, I. (2015). Olive oil pilot-production assisted by pulsed electric field: Impact on extraction yield, chemical parameters and sensory properties. *Food Chemistry, 167*, 497–502.

Puértolas, E., Saldaña, G., Álvarez, I., & Raso, J. (2010). Effect of pulsed electric field processing of red grapes on wine chromatic and phenolic characteristics during aging in oak barrels. *Journal of Agricultural and Food Chemistry, 58*(4), 2351–2357.

Qiu, X., Sharma, S., Tuhela, L., Jia, M., & Zhang, Q. H. (1998). An integrated pef pilot plant for continuous nonthermal pasteurization of fresh orange juice. *Transactions of the ASAE, 41*(4), 1069–1074.

Rechcigl, M. (1984). *Food for human use*. Boca Raton, FL: CRC Press.

Riener, J., Noci, F., Cronin, D. A., Morgan, D. J., & Lyng, J. G. (2009). Effect of high intensity pulsed electric fields on enzymes and vitamins in bovine raw milk. *International Journal of Dairy Technology, 62*(1), 1–6.

Rivas, A., Rodrigo, D., Martínez, A., Barbosa-Cánovas, G. V., & Rodrigo, M. (2006). Effect of PEF and heat pasteurization on the physical–chemical characteristics of blended orange and carrot juice. *LWT - Food Science and Technology, 39*(10), 1163–1170.

Salvia-Trujillo, L., Morales-de la Peña, M., Rojas-Graü, A., Welti-Chanes, J., & Martín-Belloso, O. (2017). Mineral and fatty acid profile of high intensity pulsed electric fields or thermally treated fruit juice-milk beverages stored under refrigeration. *Food Control, 80*, 236–243.

Sampedro, F., Rodrigo, D., Martínez, A., Barbosa-Cánovas, G. V., & Rodrigo, M. (2006). Review: Application of pulsed electric fields in egg and egg derivatives. *Food Science and Technology International, 12*(5), 397–405.

Sánchez-Moreno, C., Plaza, L., Elez-Martínez, P., De Ancos, B., Martín-Belloso, O., & Cano, M. P. (2005). Impact of high pressure and pulsed electric fields on bioactive compounds and antioxidant activity of orange juice in comparison with traditional thermal processing. *Journal of Agricultural and Food Chemistry, 53*(11), 4403–4409.

Schilling, S., Toepfl, S., Ludwig, M., Dietrich, H., Knorr, D., Neidhart, S., & Carle, R. (2008). Comparative study of juice production by pulsed electric field treatment and enzymatic maceration of apple mash. *European Food Research and Technology, 226*(6), 1389–1398.

Sepúlveda-Ahumada, D. R., Ortega-Rivas, E., & Barbosa-Cánovas, G. V. (2000). Quality aspects of cheddar cheese obtained with milk pasteurized by pulsed electric fields. *Food and Bioproducts Processing: Transactions of the Institution of of Chemical Engineers, Part C, 78*(2), 65–71.

Sharma, P., Oey, I., & Everett, D. W. (2014). Effect of pulsed electric field processing on the functional properties of bovine milk. *Trends in Food Science & Technology, 35*(2), 87–101.

Sharma, P., Oey, I., & Everett, D. W. (2015). Interfacial properties and transmission electron microscopy revealing damage to the milk fat globule system after pulsed electric field treatment. *Food Hydrocolloids, 47*, 99–107.

Sharma, P., Oey, I., & Everett, D. W. (2016). Thermal properties of milk fat, xanthine oxidase, caseins and whey proteins in pulsed electric field-treated bovine whole milk. *Food Chemistry, 207*, 34–42.

Shynkaryk, M. V., Lebovka, N. I., Lanoisellé, J.-L., Nonus, M., Bedel-Clotour, C., & Vorobiev, E. (2009). Electrically-assisted extraction of bio-products using high pressure disruption of yeast cells (*Saccharomyces cerevisiae*). *Journal of Food Engineering, 92*(2), 189–195.

Silve, A., Papachristou, I., Wüstner, R., Sträßner, R., Schirmer, M., Leber, K., & Frey, W. (2018). Extraction of lipids from wet microalga *Auxenochlorella protothecoides* using pulsed electric field treatment and ethanol-hexane blends. *Algal Research, 29*, 212–222.

Soliva-Fortuny, R., Balasa, A., Knorr, D., & Martín-Belloso, O. (2009). Effects of pulsed electric fields on bioactive compounds in foods: A review. *Trends in Food Science & Technology, 20*(11–12), 544–556.

Sui, Q., Roginski, H., Williams, R. P. W., Versteeg, C., & Wan, J. (2011). Effect of pulsed electric field and thermal treatment on the physicochemical and functional properties of whey protein isolate. *International Dairy Journal, 21*(4), 206–213.

Sulaiman, A., Farid, M., & Silva, F. V. (2017). Quality stability and sensory attributes of apple juice processed by thermosonication, pulsed electric field and thermal processing. *Food Science and Technology International, 23*(3), 265–276.

Suwandy, V., Carne, A., van de Ven, R., Bekhit, A. E.-D. A., & Hopkins, D. L. (2015a). Effect of pulsed electric field on the proteolysis of cold boned beef *M. Longissimus lumborum* and *M. Semimembranosus*. *Meat Science, 100*, 222–226.

Suwandy, V., Carne, A., van de Ven, R., Bekhit, A. E.-D. A., & Hopkins, D. L. (2015b). Effect of pulsed electric field treatment on hot-boned muscles of different potential tenderness. *Meat Science, 105*, 25–31.

Suwandy, V., Carne, A., van de Ven, R., Bekhit, A. E.-D. A., & Hopkins, D. L. (2015c). Effect of pulsed electric field treatment on the eating and keeping qualities of cold-boned beef loins: Impact of initial pH and fibre orientation. *Food and Bioprocess Technology, 8*(6), 1355–1365.

Suwandy, V., Carne, A., van de Ven, R., Bekhit, A. E.-D. A., & Hopkins, D. L. (2015d). Effect of repeated pulsed electric field treatment on the quality of cold-boned beef loins and topsides. *Food and Bioprocess Technology, 8*(6), 1218–1228.

Töpfl, S. (2006). *Pulsed Electric Fields (PEF) for permeabilization of cell Membranes in food- and bioprocessing applications, process and equipment design and cost analysis*. Berlin, Germany: Berlin University of Technology.

Töpfl, S., & Heinz, V. (2007). Application of pulsed electric fields to improve mass transfer in dry cured meat products. *Fleischwirtschaft International: Journal for Meat Production and Meat Processing, 1*, 62–64.

Torregrosa, F., Cortés, C., Esteve, M. J., & Frígola, A. (2005). Effect of high-intensity pulsed electric fields processing and conventional heat treatment on orange-carrot juice carotenoids. *Journal of Agricultural and Food Chemistry, 53*(24), 9519–9525.

Turk, M. F., Billaud, C., Vorobiev, E., & Baron, A. (2012). Continuous pulsed electric field treatment of French cider apple and juice expression on the pilot scale belt press. *Innovative Food Science & Emerging Technologies, 14*, 61–69.

Turk, M. F., Vorobiev, E., & Baron, A. (2012). Improving apple juice expression and quality by pulsed electric field on an industrial scale. *LWT - Food Science and Technology, 49*(2), 245–250.

Vallverdú-Queralt, A., Oms-Oliu, G., Odriozola-Serrano, I., Lamuela-Raventos, R. M., Martín-Belloso, O., & Elez-Martínez, P. (2012). Effects of pulsed electric fields on the bioactive compound content and antioxidant capacity of tomato fruit. *Journal of Agricultural and Food Chemistry, 60*(12), 3126–3134.

Vicaş, S. I., Bandici, L., Teuşdea, A. C., Turcin, V., Popa, D., & Bandici, G. E. (2017). The bioactive compounds, antioxidant capacity, and color intensity in must and wines derived from grapes processed by pulsed electric field. *CyTA - Journal of Food, 15*(4), 553–562.

Vorobiev, E., & Lebovka, N. (2009). Pulsed-electric fields induced effects in plant tissues: Fundamental aspects and perspectives of applications. In E. Vorobiev & N. Lebovka (Eds.), *Electrotechnologies for extraction from food plants and biomaterials* (pp. 39–81). New York, NY: Springer.

Walkling-Ribeiro, M., Noci, F., Cronin, D. A., Lyng, J. G., & Morgan, D. J. (2009). Shelf life and sensory evaluation of orange juice after exposure to thermosonication and pulsed electric fields. *Food and Bioproducts Processing, 87*(2), 102–107.

Walkling-Ribeiro, M., Noci, F., Cronin, D. A., Lyng, J. G., & Morgan, D. J. (2010). Shelf life and sensory attributes of a fruit smoothie-type beverage processed with moderate heat and pulsed electric fields. *LWT - Food Science and Technology, 43*(7), 1067–1073.

Walstra, P. (1983). Physical chemistry of milk fat globules. In P. F. Fox (Ed.), *Developments in dairy chemistry - 2* (pp. 119–158). Dordrecht, Netherlands: Springer.

Walstra, P. (1990). On the stability of casein micelles. *Journal of Dairy Science, 73*(8), 1965–1979.

Walstra, P., Wouters, J. T. M., & Geurts, T. J. (2014). *Dairy science and technology*. Boca Raton, FL: Taylor and Francis.

Wiktor, A., Nowacka, M., Dadan, M., Rybak, K., Lojkowski, W., Chudoba, T., & Witrowa-Rajchert, D. (2016). The effect of pulsed electric field on drying kinetics, color, and microstructure of carrot. *Drying Technology, 34*(11), 1286–1296.

Wu, Y., Mittal, G. S., & Griffiths, M. W. (2005). Effect of pulsed electric field on the inactivation of microorganisms in grape juices with and without antimicrobials. *Biosystems Engineering, 90*(1), 1–7.

Xiang, B. Y., Ngadi, M. O., Ochoa-Martinez, L. A., & Simpson, M. V. (2011). Pulsed electric field-induced structural modification of whey protein isolate. *Food and Bioprocess Technology, 4*(8), 1341–1348.

Xiang, B. Y., Simpson, M. V., Ngadi, M. O., & Simpson, B. K. (2011). Flow behaviour and viscosity of reconstituted skimmed milk treated with pulsed electric field. *Biosystems Engineering, 109*(3), 228–234.

Xu, S., Walkling-Ribeiro, M., Griffiths, M. W., & Corredig, M. (2015). Pulsed electric field processing preserves the antiproliferative activity of the milk fat globule membrane on colon carcinoma cells. *Journal of Dairy Science, 98*(5), 2867–2874.

Yeom, H. W., Evrendilek, G. A., Jin, Z. T., & Zhang, Q. H. (2004). Processing of yogurt-based products with pulsed electric fields: Microbial, sensory and physical evaluations. *Journal of Food Processing and Preservation, 28*(3), 161–178.

Yeom, H. W., Streaker, C. B., Zhang, Q. H., & Min, D. B. (2000). Effects of pulsed electric fields on the quality of orange juice and comparison with heat pasteurization. *Journal of Agricultural and Food Chemistry, 48*(10), 4597–4605.

Yin, Y., Han, Y., & Liu, J. (2007). A novel protecting method for visual green color in spinach puree treated by high intensity pulsed electric fields. *Journal of Food Engineering, 79*(4), 1256–1260.

Zeng, F., Gao, Q., Han, Z., Zeng, X., & Yu, S. (2016). Structural properties and digestibility of pulsed electric field treated waxy rice starch. *Food Chemistry, 194*, 1313–1319.

Zeng, X., Han, Z., & Zi, Z. (2010). Effects of pulsed electric field treatments on quality of peanut oil. *Food Control, 21*(5), 611–614.

Zhang, S., Yang, R., Zhao, W., Hua, X., Zhang, W., & Zhang, Z. (2011). Influence of pulsed electric field treatments on the volatile compounds of milk in comparison with pasteurized processing. *Journal of Food Science, 76*(1), C127–C132.

Zhao, W., Yang, R., Tang, Y., & Lu, R. (2007). Combined effects of heat and PEF on microbial inactivation and quality of liquid egg whites. *International Journal of Food Engineering, 3*(4). https://doi.org/10.2202/1556-3758.1256

Zulueta, A., Esteve, M. J., Frasquet, I., & Frígola, A. (2007). Fatty acid profile changes during orange juice-milk beverage processing by high-pulsed electric field. *European Journal of Lipid Science and Technology, 109*(1), 25–31.

Chapter 3
Impact of Ultrasound on Food Constituents

Mladen Brnčić and Jana Šic Žlabur

3.1 Introduction

Ultrasound is one of the non-invasive technologies, which successfully find widespread use in numerous processes in food technology. It represents one of the novel technologies that in a very short time rapidly found evolution and implementation in various food industry processes and commercial products. Some of the mentioned food processes in which ultrasound finds its application include drying, freezing, homogenization, sterilization, extraction, bleaching, crystallization, emulsification, and filtration. Specific equipment required in mentioned food industry applications is constructed to fit ultrasound principles and nowadays successfully applied even at the level of larger capacity and industrial scale. All mentioned prove that ultrasound is successfully implemented and commercialized in the food industry and for the food products such as fruits and vegetables (dried, juices), meat, and dairy products (milk, cheese, chocolate) (Chemat, Zill-e-Huma, & Khan, 2011; Kiani, Sun, & Zhang, 2013; Tao, García, & Sun, 2014; Zinoviadou et al., 2015).

M. Brnčić (✉)
Faculty of Agriculture, University of Zagreb, Zagreb, Croatia
e-mail: mbrncic@pbf.hr

J. Šic Žlabur
Faculty of Food Technology and Biotechnology, University of Zagreb, Zagreb, Croatia

S. Roohinejad et al. (eds.), *Effect of Emerging Processing Methods on the Food Quality*,
https://doi.org/10.1007/978-3-030-18191-8_3

3.2 Advantageous and Challenges of Using Ultrasound on Food Quality

Ultrasound technology shows a significant positive impact on the physicochemical and nutritional characteristics of various foods and following numerous advantages in its application: significantly shortened duration of the process. Full food process can be completed in a very short time (seconds or minutes) compared to the conventional methods with easy and safe use. In addition, the equipment is mostly constructed without moving parts and without any danger for the operator. Furthermore, for example, in extraction processes, ultrasound treatment significantly enhances the yield of various chemical compounds. Compared with traditional technologies, it preserves food ingredients such as vitamins, pigments, phenols, and most of the thermo-labile compounds. Moreover, ultrasound represents environmentally friendly technology coupled with the significant energy savings and also avoidance of hazardous organic solvents usage (Table 3.1) (Chemat et al., 2011).

Table 3.1 Possibilities and advantages of ultrasound application in food processing compared to conventional methods (Chemat et al., 2011)

Product type	Application	Conventional method	Ultrasound advantages
Meat Fruits and vegetables, dairy products	Freezing/ crystallization	Cooling, freezing by contact, by immersion	Less time
			Small crystals
			Improving diffusion
			Rapid temperature decreasing
Meat Vegetables	Cooking	Stove, fryer	Less time
			Improving heat transfer
			Organoleptic quality
Fruits and vegetables	Drying	Atomization, hot gas stream, freezing, pulverization	Less time
			Improving organoleptic quality
			Improving heat transfer
Juices (liquids)	Filtration	Filters (membranes semi-permeable)	Less time
			Improving the filtration rate
Food emulsions (ketchup, mayonnaise)	Emulsification	Mechanical treatment	Less time
			Emulsion stability increased
Alcoholic beverages (wine)	Oxidation	Contact with air	Less time

3.2.1 *Nutritional Properties*

3.2.1.1 Fruits and Vegetables

Ultrasound is categorized as a non-thermal method with an average increase of temperature between 1 and 2 °C/min (directly immersed sonotrode). This represents one of the main ultrasound advantages in preserving various nutritional properties in the processing of food and vegetables. Thermo-labile compounds such as vitamins, pigments, phenols, and in general, bioactive compounds are sensitive even to a mild temperature increase, which is by using ultrasound significantly reduced (Aadil, Zeng, Han, & Sun, 2013; Pérez-Grijalba et al., 2017; Sulaiman, Farid, & Silva, 2017). Aadil et al. (2013) investigated the influence of sonication on some physicochemical parameters of grapefruit juice: acidity and total soluble solids content. Sonication treatments were performed immediately after juice extraction in an ultrasonic bath at 28 kHz frequency, a power of device set at 70% and a temperature of 20 °C. Sonication of juices lasted for 30, 60, and 90 min. Compared to the control sample of grapefruit juice, pH value, acidity and total soluble solids did not change, even after sonication treatments for 60 and 90 min. Similar results of sonication effect on some physicochemical parameters (e.g. acidity and total soluble solids) were obtained by various research studies (Adekunte, Tiwari, Cullen, Scannell, & O'Donnell, 2010; Bhat, Kamaruddin, Min-Tze, & Karim, 2011; Tiwari, Muthukumarappan, O'Donnell, & Cullen, 2008; Zou, Hou, Zou, & Hou, 2017; Zou, Jiang, Zou, & Jiang, 2016). Opposite results were obtained by Šic Žlabur et al. (2017) in whose research some changes in the content of total acids and total soluble solids were determined during the sonication of apple juice with added chokeberry powder. Compared to the juice samples conventionally treated and sonicated ones, it can be concluded that major changes in the mentioned physicochemical parameters did not occur. The main mechanism of high-intensity ultrasound is the phenomenon of transient cavitation, which results in a mild temperature increase of the system. The temperature increase affects some physical properties of the liquid, primarily in systems where the applied ultrasound power (amplitude) is higher, for example in systems with direct immerse ultrasound probe (Knorr, Zenker, Heinz, & Lee, 2004; Šic Žlabur et al., 2015). Except for the described parameters, some studies researched the effect of sonication on some physical parameters of liquid samples, such as electrical conductivity, density, and viscosity. Aadil et al. (2015) investigated the influence of power ultrasound on the electrical conductivity of grapefruit juice. In general, liquid foods such as fruit and vegetable juices are good electrical conductors due to the presence of water, vitamins, minerals, and proteins, which are good conductors (Aadil et al., 2015; Abid et al., 2013; Zinoviadou et al., 2015). Therefore, higher values of electrical conductivity suggest a greater content of mentioned compounds. Some studies proved a positive impact of sonication on electrical conductivity values. Aadil et al. (2015) reported an increase of electrical conductivity in samples treated by ultrasound during 90 min for 11% compared to the control (untreated) sample. In addition, a slight increase was determined during sonication of blueberry juice and carrot juice, in research works conducted by Zou et al. (2017) and Zou et al. (2016), respectively. A positive effect of sonication on the electrical conductivity of liquids could be related to the ultrasound possibility to break up the cell walls and release (isolation) of cell nutrients (e.g., vitamins, minerals, proteins, etc.) in solution. Another physical characteristic of fluid under the influence of ultrasound is viscosity, whereby sonication can affect its increase or decrease, depending on the fluid type. In liquids, such as fruit juices, the sugar content is relatively high which ultimately can influence the viscosity during sonication. Sonication influenced the isolation, permeation of sugar molecules from cell membranes into solution, which resulted in an increase of viscosity of blueberry juice, first of all, because of sugar concentration in the colloidal solution is correlated to viscosity (Suárez-Jacobo et al., 2011). Ultrasound treatment also shows positive effects on various biologically active compounds specifically on the nutritional composition of fruits and vegetables such as vitamins, minerals, phenols, and pigments. A significant increase in vitamin C content for even 28.45% in grapefruit juice was reported (Aadil et al., 2013). Šic Žlabur et al. (2017) showed also an increase in the content of vitamin C during sonication treatment. In the mentioned research, apple juice with added chokeberry powder was sonicated in an ultrasonic bath (35 kHz frequency, 140 W) for 5, 10, 15, 20, 25, and 30 min, which significantly contributed to the increase of the vitamin C yield. Increase for even 49% in vitamin C content was recorded compared to the sample classically treated for the same period of time. Vitamin C increase was also recorded in orange juice treated by ultrasound in a system with an ultrasonic processor of 1500 W and 20 kHz, and a probe of 19 mm diameter (Tiwari, O'Donnell, & Cullen, 2009). Moreover, Abid et al. (2013) reported an increase in vitamin C content for 34% during sonication of apple juice in an ultrasonic bath for 90 min. Some research studies noticed a possibility of slight decrease of vitamin C content during sonication treatment that can be caused by cavitation mechanism (Adekunte et al., 2010; Dias et al., 2015; Ordóñez-Santos, Martínez-Girón, & Arias-Jaramillo, 2017). During the sonolysis (action of ultrasonic waves that act to break down or decompose a substance) of water molecules present in juice sample, hydrogen ions (H^+), free radicals (O^-, OH^-, HO_2^-) and hydrogen peroxide (H_2O_2) are formed (Pétrier, Combet, & Mason, 2007). Interactions of free radicals promoted during sonication are related to the occurrence of the oxidation process, which can be caused by vitamin C degradation. Compounds such as flavonoids, phenolic acids, flavones, flavonols, and in general antioxidants, show a positive impact during sonication treatment mainly due to the formation of OH^- radicals, which can improve the functionality of the mentioned compounds. In other words, phenol activity in foods depends, among other factors, on the hydroxylation degree, so as the hydroxylation degree is higher, the functionality of antioxidants will be improved (Ashokkumar et al., 2008). Various scientific studies proved that sonication is correlated with the increase of phenolic compounds in different fruit and vegetable samples. Bhat et al. (2011) noted an increase of total phenol compounds in sonicated lime juice for 28% compared to the control sample, while Aadil et al. (2013) noticed an increase of 9% in grapefruit juice. Furthermore, Alighourchi, Barzegar, Sahari, and Abbasi (2013) noted an increase of total phenolic compounds content during sonication of pomegranate juice during different applied ultrasonic powers for 17%. Similarly, Abid et al. (2013) showed an increase in total phenol content of sonicated apple juice for 9.5%, total flavonoid content for 30% and total flavonols for 44% compared to the control sample. Moreover, Zou et al. (2017) observed the same behavior in sonicated blueberry juice as an increase of 11%, whereas Šic Žlabur et al. (2017) observed an increase in total phenol content for 12% in apple juice with added chokeberry powder. In addition, a significant increase of total phenol content, even up to 31%, was determined in sonication of different stevia (*Stevia rebaudina* Bertoni) extracts (Šic Žlabur et al., 2015) and during extraction of phenolic compounds from Chilean papaya fruits for 11% (Uribe, Delgadillo, Giovagnoli-Vicuña, Quispe-Fuentes, & Zura-Bravo, 2015). Sonication treatment and precisely cavitation mechanism cause the cell wall disruption that facilitates the release of bound and free phenolic compounds. Except mentioned, the second possible reason for the enhancement of phenolic compounds content during sonication might be an increase of polyphenol oxidase (PPO) enzyme activity (Bhat et al., 2011). In Table 3.2, some examples of polyphenol extraction possibilities by ultrasound from various vegetable sources are presented. From the group of polyphenols, it is important to emphasize the effect of sonication treatment on anthocyanin level, since some research studies suggested a negative impact on anthocyanin content (Tiwari, O'Donnell, Muthukumarappan, & Cullen, 2008; Tiwari, O'Donnell, Patras, & Cullen, 2008). On the other hand, Šic Žlabur et al. (2017) noted a positive impact of sonication on anthocyanin content in apple juice with added chokeberry powder. In the mentioned research, sonication has contributed to an increase in anthocyanin yield for even 72%. Also, a positive effect of sonication treatment is recorded for the extraction of anthocyanins from "Purple Majesty" potato (Mane et al., 2015) and from dried black chokeberry (Galvan d'Alessandro, Kriaa, Nikov, & Dimitrov, 2012). In a similar work, Zou et al. (2017) recorded an increase of

Table 3.2 Ultrasound possibilities of polyphenol extraction from various fruit and vegetable sources (Medina-Torres, Ayora-Talavera, Espinosa-Andrews, Sánchez-Contreras, & Pacheco, 2017)

Vegetable source	Compounds	Advantages	Reference
Pomegranate (peel)	Total polyphenols	Increased yield (24%)	Pan, Qu, Ma, Atungulu, and McHugh (2011)
Grapes (seeds)	Phenols	Enhanced extraction yield	Soria and Villamiel (2010)
Grapes (fruit)	Resveratrol	Increased yield (24–28%)	Barba, Zhu, Koubaa, Sant'Ana, and Orlien (2016)
Grapes by-products	Polyphenols	Increased yield (50%)	Barba et al. (2016)
Garlic, wild	Bioactive compounds	Better extraction	Tomšik et al. (2016)
Grape, pomace	Polyphenols	Increased extraction	Drosou, Kyriakopoulou, Bimpilas, Tsimogiannis, and Krokida (2015)
Purple Majesty potato	Anthocyanins	Increased extraction	Mane, Bremner, Tziboula-Clarke, and Lemos (2015)

anthocyanidin content in sonicated blueberry juice for 17%. In a research conducted by Alighourchi et al. (2013), a slight increase (4%) of total monomeric anthocyanins in pomegranate juice was recorded while lower ultrasonic power (50%) was applied, however, when higher ultrasonic powers (75% and 100%) were applied, a decrease of anthocyanins was observed. Anthocyanins are thermolabile compounds, which the content strongly depends on various environmental factors such as light, temperature, pH value, oxygen presence, etc. During sonication, the optimization of process factors is crucial for anthocyanin preservation (Tiwari, Patras, Brunton, Cullen, & O'Donnell, 2010), mainly because ultrasound factors such as higher ultrasonic power or longer time period of sonication can strongly affect the anthocyanin degradation.

Besides polyphenols, yet other important nutritional constituents of fruits and vegetables are plant pigments, which give fruits, vegetables, and their products characteristic colors. In fruits, vegetables and their products, the most common pigments are chlorophylls, carotenoids, betalains, and anthocyanins. Mentioned pigments are extremely sensitive compounds especially in terms of food processing. Food processing operations, which mostly involve the use of heat, cause significant degradation and loss of plant pigments (Ngamwonglumlert, Devahastin, & Chiewchan, 2017). Different research studies show a positive impact of sonication on plant pigments content. From the carotenoids, an increase of the total carotenoid content has been published by Ordóñez-Santos et al. (2017) in sonication of Cape gooseberry (*Physalis peruviana* L.) juice; for 14% in sonication of carrot juice (Zou et al., 2016), while an increase of 13% in β-carotene content was recorded in sonicated pomegranate juice (Alighourchi et al., 2013). Moreover, Lianfu and Zelong (2008) have proven an increase of lycopene yield in the application of ultrasound combined with microwave-assisted extraction from tomato paste. Rosu, Nistor, Miron, Popa, and Cojocaru (2017) studied the influence of ultrasound-assisted extraction of photosynthetic pigments, among others on the content of chlorophyll a and b from a dried drill. Ultrasound-assisted extraction was significantly more efficient compared to the conventional technique, with an increase of chlorophyll a yield by 10%, and chlorophyll b by 20%.

Since sonication is extremely effective in increasing the yield of different polyphenolic compounds, antioxidant activity is also under its strong influence. In general, antioxidant activity is in positive correlation with the content of vitamins, minerals, phenolic compounds, pigments, and in general bioactive compounds. Therefore, samples with higher content of bioactive compounds will exhibit a stronger and higher antioxidant activity (Carbonell-Capella, Barba, Esteve, & Frígola, 2014; Šic Žlabur et al., 2015). In addition, different research studies proved the positive effect of ultrasound on the antioxidant activity of different fruit and vegetable samples. Sulaiman et al. (2017), in the thermosonication of apple juice, showed 20% higher values of antioxidant activity compared to the untreated juice, Aadil et al. (2015) showed about 11% increase of antioxidant activity in sonicated grapefruit juice, while Bhat et al. (2011) showed an increase in the antioxidant activity of sonicated lime juice for 35%. In addition, Abid et al. (2013) observed an increase of 21% in antioxidant activity of sonicated apple juice, whereas Zou et al. (2017) noticed an increase of antioxidant activity in sonicated blueberry juice for even 38%.

3.2.1.2 Meat Products

One of the most widespread procedures in meat processing is brining, which is primarily used to maintain the juiciness, tenderness, flavors and shelf life of meat products (Ozuna, Puig, García-Pérez, Mulet, & Cárcel, 2013). Brine represents a mixture of salt and water in which meat is marinated for a certain time period. Nowadays, we are faced with many chronic illnesses associated with excessive salt intake, including most of all high blood pressure and the risk of coronary diseases (Aburto et al., 2013). Precisely, because of the higher rates of chronic diseases associated with excessive salt intake. World Health Organization in 2012 recommended a reduction of salt intake for adults from 5 g/day to 2 g/day (Lim et al., 2012). At the same time, different consumer habits from Europe, North America, and Australia suggest that even 20% salt in the diet come from meat products (Ruusunen & Puolanne, 2005). Therefore, one of the biggest challenges facing the meat processing industry is to create products with less salt. Ultrasound technology is successfully implemented to reduce salt content but simultaneously speeding up salting and curing processes in meat production (Inguglia, Zhang, Tiwari, Kerry, & Burgess, 2017). Different research studies proved significantly faster NaCl increase in meat treated by ultrasound (Cárcel, Benedito, Bon, & Mulet, 2007; Siró et al., 2009). The cavitation phenomenon consequently affecting the mass transfer between the liquid and solid matrices, which is the base for further processes developing on the tissue level and influencing in general salt reduction. In the process of meat brining, ultrasound helps better salt distribution causing increased salt gain rate, which ultimately results in a lower need of NaCl used in the brine solution (Alarcon-Rojo, Janacua, Rodriguez, Paniwnyk, & Mason, 2015; Tao & Sun, 2015). Except for brining, one of the most used procedures in the preparation of various meat products in the meat industry is marination. Similar to brining, marination process includes incorporation of an aqueous or oily solution, which contains different ingredients and/or additives usually salt, polyphosphates, flavorings, etc., in different types of meat muscle (e.g., pork, beef, chicken, turkey, lamb, etc.) (Tapasco, Restrepo, & Suarez, 2011). Besides extension of the shelf life of such prepared meat product, one of the most important functions of marination is to increase the meat juiciness and to positively affect the meat texture by increasing the water retention capacity in myofibrillar tissue (Xargayó, Lagares, Fernández, Borrell, & Juncá, 2004). Several techniques for marination are often in use (e.g., injection, immersion and massaging) (Alvarado & McKee, 2007) while beside mentioned traditional/conventional techniques, alternative/non-thermal ones such as ultrasound-assisted marination technology, are increasingly developed (Chemat et al., 2011). One of the successful applications of ultrasound in the marination process was demonstrated in the research from the group of authors González-González et al. (2017). From bovine muscle (*Longissimus dorsi*) samples prior to treatments, bone, fat, and connective tissue were removed and meat samples were cut into pieces having dimensions of 2 × 2 × 2.5 cm. Such meat samples were placed in polypropylene bags and brine solution prepared from a mixture of different ingredients and additives (sodium chloride, monosodium glutamate, garlic, onion, black pepper, dextrose, citric acid, silicon dioxide, and yellow 5 (tartrazine)) and water was added. Sonication treatments were performed in an ultrasonic bath at a frequency of 40 kHz and an intensity of 11 W/cm^2 while the total time for marination was 20, 40 and 60 min. Based on the obtained results, the authors concluded that ultrasonic treatment can be a good alternative to traditional marination process due to the better distribution of brine solutes in the meat. However, the results of the sensorial analysis showed that ultrasound-assisted marination did not significantly affect the sensory properties of beef. Triglycerides, fats, and more specifically fatty acids are an important part of daily diet, especially in the term of energetic value. However, lately, more and more foods with increased fat content are consumed, which ultimately result in increased blood fat and consequently in the development of various chronic diseases (e.g., cardiovascular diseases) (Troy, Tiwari, & Joo, 2016). Among others, pork is a very popular meat type in daily consumption, which besides valuable compounds such as proteins, fibers, vitamins, and minerals, contains high lipid levels (Ojha et al., 2017). Many authors suggest that the nutritional value of meat and meat products can be increased by reducing the total fat content and energetic value, and by enhancing the fatty acid content (Jiménez-Colmenero, 2013; Olmedilla-Alonso, Jiménez-Colmenero, & Sánchez-Muniz, 2013). In the context of nutritive value, it is not enough to increase the total amount of fatty acids but to reduce the content of saturated fatty acids and at the same time to increase the content of unsaturated fatty acids, which shows numerous health benefits (Mapiye et al., 2015). One of the possibilities to enhance the fatty acid profile of meat type with a high content of saturated fatty acids (i.e., pork) is the encapsulation of omega-3 fatty acids (unsaturated) into the meat. From previous research studies, fish oil, rich in omega-3 fatty acids were successfully microencapsulated in order to enhance the profile of fatty acids. So far, several techniques have been successfully implemented for needs of fatty acids encapsulation. However, ultrasound has shown a number of beneficial effects, especially in the terms of assisted diffusion of various ingredients, among others, fatty acids, between food matrices (Cárcel et al., 2007; Ozuna et al., 2013). Ojha et al. (2017) investigated ultrasonic-assisted incorporation of nano-encapsulated omega-3 fatty acids in pork meat. More precisely, for research purposes, porcine meat of *M. semitendinosus* was sliced into meat cubes (dimensions 4 × 4 × 4 mm) while all visible fat from each muscle was manually removed. Such prepared pork meat cubes were immersed in a suspension of nanovesicles where fish oil was used and placed in an ultrasonic bath (frequency 25 kHz) for 30 and 60 min. Mentioned authors proved significant positive impact of ultrasound in combination with nanoencapsulation on increase of beneficial fatty acids in pork meat, primarily the increase of eicosapentaenoic acid (EPA) and docosahexaenoic acid (DHA). In addition, it is important to emphasize that the authors highlighted that the atherogenic index (AI) remained constant after treatment, while the thrombogenic index (TI) and omega-3/omega-6 ratio were beneficially modified.

3.2.1.3 Dairy Products

High power ultrasound has been used in various milk processing applications such as acceleration of mass transport processes in mixing, degassing of liquid foods, filtration, polysaccharide depolymerization, viscosity modification, induction of oxidation/reduction reactions, extraction of proteins, microbial and enzyme inactivation, induction of nucleation for crystallization and preparation of submicron emulsions (Arzeni et al., 2012). It may cause two different effects: physical and chemical effects. Ultrasound could be used in liquid and gaseous media. In liquid medium, ultrasound mechanism is based on the cavitational phenomenon. High-intensity ultrasound is characterized by low frequencies within ultrasonic spectra of sound and range from 18 to 100 kHz. Well coupled with strong nominal output power (W), it can cause a huge release of energy, making both splitting of intermolecular and intramolecular bonds. During such activity in the liquid medium, cavitation bubbles are formed in microscale size filled with gasses. Bubble collapses once reaching a critical size, spreading energy into surrounding liquid medium. Consequences of such ultrasonic mechanism could lead to temperatures up to 5500 K and pressure up to 100 MPa (Dujmić et al., 2013). The functional properties of starch, proteins and other food constituents have been significantly affected by the physical and chemical effects generated by acoustic cavitation, shear, micro-jet and acoustic streaming (Chandrapala, Oliver, Kentish, & Ashokkumar, 2012). Pasteurized homogenized milk (PHSM) was the issue of the work of Shanmugam, Chandrapala, and Ashokkumar (2012) where samples were treated with ultrasound of 20 kHz frequency, 20 and 41 W of actual dispersed device power levels, and varied times of treatment. They concluded that whey proteins, as well as whey-whey aggregates that were present in PHSM as constituents during aggressive treatment of high-intensity ultrasound, were denatured from their original form and shape, which results in the formation of whey-whey/whey-casein aggregates. Native-PAGE has shown that ultrasound treatment significantly increases κ-casein content from whey protein when applying up to 45 min of treatment for both actual power levels propagated within the samples (20/41 W). Further treatment (up to 60 min) led to decrease the κ-casein content. Changes in size of fat globules and casein micelles were significant for both power levels of ultrasonic propagation, while milk viscosity did not show any change after ultrasound treatment under any applied conditions. Chandrapala, Zisu, Kentish, and Ashokkumar (2012) evaluated the changes in the secondary and tertiary structure of α-lactalbumin, β-lactoglobulin, and their mixtures after high-intensity ultrasound (HIU) treatment. Sonication of β-lactoglobulin showed a slight increase in the reactive thiol content and surface hydrophobicity due to the time of treatment and unfolding of the dimer structure. Only minor changes in the secondary and tertiary structures occurred for individual β-lactoglobulin. α-Lactalbumin was more influenced by HIU treatment. Mixtures of α-lactalbumin/β-lactoglobulin upon sonication have shown an increase of hydrophobicity, but only up to the first 10 min of treatment, while prolonged ultrasonic treatment led to decrease the hydrophobicity followed by a significant increase of the aggregate particle size. This fact probably occurred due to exposed thiol in the β-lactoglobulin ability to create interactions with α-lactalbumin disulfide bonds. Chandrapala, Zisu, Palmer, Kentish, and Ashokkumar (2011) also examined the structural changes in milk globular proteins, α-lactalbumin and β-lactoglobulin in whey protein concentrate. Whey proteins and whey protein concentrates are important ingredients of dairy industry, primarily due to their valuable nutritional composition (Brnčić, Ježek, Rimac Brnčić, Bosiljkov, & Tripalo, 2008; Brnčić, Tripalo, Brncic, et al., 2009; Brnčić, Tripalo, Penava, et al., 2009; Brnčić, et al., 2011). Since one of the necessary steps in the dairy industry for the production of different products are processes assuming the use of heat, one of the biggest challenges of the dairy industry is the reduction of the processes involving the application of temperatures higher than 70 °C. Denaturation of proteins, as well as the exposure of internal thiol groups, begins at temperatures greater than 70 °C (Bernal & Jelen, 1985). The main consequence of whey protein denaturation is the formation of large aggregates composed of complexes formed from denaturated whey proteins themselves and casein micelles, which ultimately reflects on excessive thickening or gelling during thermal processing (Paulsson & Dejmek, 1990). Because of all mentioned, greater development of methods that do not assume the use of high temperatures is necessary. Chandrapala et al. (2011) investigated the effect of sonication on reconstituted whey protein concentrate solutions. Precisely, whey protein concentrate solutions were sonicated in a system with an ultrasonic probe of 19 mm diameter, a frequency of 20 kHz, a nominal output power of 450 W (31 W was the power delivered to the solution), an amplitude of 50%, and for 1, 5, 10, 20, 30, and 60 min. Thermal analysis of whey proteins, surface hydrophobicity, and thiol group determination was examined. Based on the conducted research, the authors concluded that sonication resulted in small changes in the thermal behavior of proteins, during which denaturation temperatures of proteins did not change. In sonicated whey protein concentrate sample solutions, changes occurred in thiol group content, and only minor changes were observed in protein secondary structures and surface hydrophobicity. Changes in protein conformation could be related to the functional properties of proteins such as solubility, foaming capacity, and flexibility. Significant changes in viscosity of protein solutions were determined during sonication treatment, where viscosity is significantly reduced (Ashokkumar, Kentish, et al., 2009; Ashokkumar, Lee, et al., 2009; Hu et al., 2013; Zisu, Bhaskaracharya, Kentish, & Ashokkumar, 2010). Reduction of viscosity during sonication is a direct consequence of cavitation, which causes a reduction in aggregate size. A significant effect of ultrasound was also observed for casein proteins. Liu, Juliano, Williams, Niere, and Augustin (2014a), Madadlou, Mousavi, Emamdjomeh, Ehsani, and Sheehan (2009) and Nguyen and Anema (2010) concluded that the average size of casein micelles was significantly reduced during sonication, which can affect its functionality. Opposite, other authors observed that sonication did not affect casein micelle size or composition, however, controlled application of ultrasonic energy can break up large casein aggregates, which influences some physical properties (e.g., viscosity) (Chandrapala, Martin, Zisu, Kentish, &

Ashokkumar, 2012). It has been shown that disruption of casein micelles is under a strong influence of sonication at high pH values (Liu et al., 2014a). In the mentioned research, reconstituted skim milk was sonicated at an energy input of 286 kJ/kg using 20 kHz at pH values ranging from 6.7 to 8. The authors concluded that at higher pH values, ultrasound caused greater disruption of casein micelles causing the release of proteins from the micellar to the serum phase.

3.2.2 Textural Properties

3.2.2.1 Fruits and Vegetables

Textural properties are a critical quality attribute of fresh fruits, vegetables, and their products in the consumer demands. For example, some fruit and vegetable species are characterized by soft texture and low firmness (high water content) and are mostly exposed during storage to a rapid loss of firmness. The direct consequence is a significant shortening of the postharvest life. Some researchers investigated the effect of ultrasound on fruit firmness (Brnčić et al., 2010). Cao et al. (2010) immersed freshly harvested strawberries in an ultrasonic water bath and treated it with different ultrasonic frequencies, at a power of 350 W, and for 10 min. Ultrasonic treatment significantly inhibited fruit softening and maintained high levels of fruit firmness during storage. Thus, it can be concluded that ultrasound positively influenced the maintaining of fruit firmness and prolonged the shelf life. Similarly, Alexandre, Brandão, and Silva (2012) observed that ultrasound significantly influenced the firmness retention of strawberry samples.

Besides, in a fresh state, fruits and vegetables are often processed, while one of the most common processes is drying. Ultrasound finds its successful application as a pre-treatment in the drying process of different fruits and vegetables. The main ultrasound advantage in the drying process is the increase of mass transfer rate between the cell and its extracellular surroundings, which results in significantly shortened drying time but also in the preservation of nutritionally valuable compounds (Fernandes & Rodrigues, 2007; Nowacka, Wiktor, Śledź, Jurek, & Witrowa-Rajchert, 2012). The changes of textural properties during ultrasound application are unavoidable since sonication causes the creation of microscopic channels in plant tissue that reduce the diffusion boundary layer and increase the convective mass transfer (Fernandes & Rodrigues, 2007; Mieszczakowska-Frąc, Dyki, & Konopacka, 2016). Nowacka et al. (2012) investigated the application of ultrasound as a pre-treatment of apple drying and its influence of some physical properties of dried material. Ultrasonic-treated apple cubes showed approximately 10% higher shrinkage, 6–20% lower density, and even of 9–14% higher porosity compared to the untreated samples. Moreover, in a study conducted by Dujmić et al. (2013), significant changes in the textural properties, specifically firmness and elasticity, of dried pear were noted while pre-treated by high-intensity ultrasound.

3.2.2.2 Meat Products

One of the greatest demands to maintain the meat quality is improving the meat tenderness. For instance, proteases, which are responsible for proteolysis (hydrolysis of peptide bonds) of myofibrillar and cytoskeletal proteins and the loss of collagen (connective tissue compound), are the main factors that affect meat tenderness (Lawrie & Ledward, 2006). In the meat processing industry, several methods (e.g., mechanical, biochemical, enzymatic, chemical, etc.) are focused on increasing the meat quality in term of achieving the satisfactory tenderness (Istrati, 2008). One of the most applicable methods to achieve the desired tenderness of the meat is enzyme (e.g., papain, bromelain, and ficin) use. However, ultrasound development significantly influenced the possibilities of improving meat quality. As a non-thermal method, ultrasound finds its application in the meat industry in terms of improving some of the most important meat properties. The main mechanism of high-intensity ultrasound action is related to the cavitation phenomenon, which causes physical disruption of the muscular tissue and results in increased release of cathepsin, calcium, and in general, results in the migration of proteins, minerals, and other components associated to meat tenderness (Turantaş, Kılıç, & Kılıç, 2015). Barekat and Soltanizadeh (2017) investigated the impact of high-intensity ultrasound and papain application on meat tenderness on samples of *Longissimus lumborum* muscles. Ultrasound treatment involved the application of ultrasonic radiation applied using a probe with the following operation conditions: frequency of 20 kHz, powers of 100 and 300 W, a constant amplitude of 100%, and time periods of 10, 20, and 30 min. Results from the mentioned research proved exceptional efficiency of ultrasound at 100 W for 20 min in combination with papain on the beef tenderness. Furthermore, other scientific studies related to the impact of ultrasonic treatment on the meat tenderness have published many results mainly related to the fact that ultrasound effects are directly correlated to the acoustic parameters: intensity, frequency, time and temperature (Stagni & De Bernard, 1968). For example, at low ultrasonic intensities (2 W/cm or 1.55 W/cm), no changes on the histological structure of muscle, and on tenderization of meat were reported (Pohlman, Dikeman, & Kropf, 1997; Zayatas, 1971). Similarly, in another research conducted by Pohlman et al. (1997), ultrasonic treatment did not significantly affect the meat properties; specifically, no improvement in aging, shear, sensory, and cooking characteristics of the beef *Pectoralis* muscle have been observed. Reported results suggested that low ultrasonic intensities do not have a significant influence on textural meat properties (not sufficient to cause myofibrillar or cellular disruption). Contrary, some other results proved an effective impact of sonication with a significant improvement of tenderness, collagen solubility, myofibrillar degradation and a significant decrease in shear force values (Dickens, Lyon, & Wilson, 1991; Nishihara & Doty, 1958; Roncalés, Ceña, Beltrán, & Jaime, 1993; Smith, Cannon, Novakofski, McKeith, & O'Brien, 1991; Stagni & De Bernard, 1968).

Collagen as a main structural protein in the extracellular space of the connective animal tissues is also under the strong influence of sonication phenomenon. Reduction in molecular weight of fibrous proteins such as collagen and keratin was observed under the mechanical effects of ultrasonic cavitation, which caused mechanical cleavage of the mentioned proteins (Coakley & Nyborg, 1978). Also, other studies recorded significant fragmentation of collagen macromolecules during sonication (Lyng, Allen, & McKenna, 1997, 1998; Nishihara & Doty, 1958). Some other studies also investigated the effect of high-intensity ultrasound on some textural properties of meat, such as the influence of sonication on the stability of the connective tissues. Roberts (1991) applied high power ultrasound on beef *Longissimus* muscle at an intensity of 2 W/cm, a frequency of 40 kHz, and during 2 h, which resulted in the improvement of the meat texture by reduction of intramuscular connective tissues level. Molecular degradation occurring during sonication and thus degradation of macromolecules such as proteins might also be the result of some sonochemical influences that occur when gas bubbles collapse and temperature in "hot spots" (localized heat increase) reaches approximately 9700 °C. Temperature increase produces chemical changes such as the formation of free radicals, which cause further destructive changes (e.g., hydroxy radicals, which are breaking hydrogen bridges) (Alligar, 1975). Except for collagen, the ultrasound effect is also proved over the myofibrillar tissues through its influence on the structure of proteins (i.e., depolymerization). Roncalés et al. (1993), Stagni and De Bernard (1968), and Zayas and Strokova (1972) in their studies proved the mechanical and chemical effects of sonication on the degradation of myofibrillar tissues in different meat types, as well as changes in protein properties, such as structure and enzymatic activity. Except for tenderness, one of the most important meat attributes, which is also linked to the meat textural and sensorial characteristics during consumption, is water-holding capacity (Huff-Lonergan & Lonergan, 2005). Water-holding capacity is defined as the ability of meat (*postmortem* muscle) to retain inherent water. It's an important property of fresh meat that affects the overall quality of the end product and influences the processing characteristics. For example, lower meat quality might result after processing in low water-holding capacity compared to fresh meat (Huff-Lonergan, 2010). Water-holding capacity is significantly affected by changes in myofibrils followed by degradation of mentioned muscle fibers during *postmortem* period. In addition, different biochemical and biophysical processes during *postmortem* period contribute to the development of the water-holding capacity of meat products (Dolatowski & Stadnik, 2007; Huff-Lonergan, 2010). Water-holding capacity of meat can be affected and improved by some technological processes such as curing, which is a very important process for meat products, primarily in term of enhanced extraction of myosin proteins that reduce hardness of meat via increasing of water holding capacity (Kang, Gao, Ge, Zhou, & Zhang, 2017). Different curing methods, among others, assume that salt injection or tumbling ensure rapid penetration of brine into muscles (Casiraghi, Alamprese, & Pompei, 2007), and very often ends with a significant mechanical damage of meat. This mechanical damage can cause further undesirable changes such as microbial contamination and lose of salt-soluble proteins (Wang, Xu, Kang, Shen, & Zhang, 2016). Another major drawback of the commonly mentioned methods is prolonged time since long duration for example in salt tumbling is required for the effectiveness of the process, which may result in increased heat and affect the product's quality (Siró et al., 2009). Considering all of the above mentioned, ultrasound is an emerging technology with very successful features in accelerating the mass transport phenomenon and inducing textural changes in meat products (Kuijpers, Kemmere, & Keurentjes, 2002). Dolatowski and Stadnik (2007) proved that ultrasound positively affected the water-holding capacity during aging of muscle samples (*M. semimembranosus*) of young bulls treated with a frequency of 45 kHz and low-intensity ultrasound (2 W/cm) for 120 s. Kang et al. (2017) investigated the influence of ultrasound during curing of beef. According to the obtained results, the authors revealed that ultrasound-assisted curing results in significant improvement in the water-holding capacity and tenderness of the meat. The samples treated by ultrasound during curing showed increased water-holding capacity values. Similar results were obtained by Stadnik, Dolatowski, and Baranowska (2008) on beef samples.

3.2.2.3 Dairy Products

Liu, Juliano, Williams, Niere, and Augustin (2014b) evaluated the gelation properties of rennet gels made from milk sonicated at different pH values (6.7 and 8.0). The results of the study showed that the gelation properties were significantly modified. Gelation was faster in rennet gels made from sonicated milk at pH 8.0 and re-adjusted back to pH 6.7, compared with milk sonicated at pH 6.7. However, at the same time, rennet gels sonicated at pH 6.7 were firmer compared to non-sonicated milk. Similar results of renneting behavior during sonication at different pH values were also determined in the works described by Chandrapala, Zisu, Kentish, and Ashokkumar (2013) and Chandrapala et al. (2013). Finally, many scientific reports mentioned that sonication exhibits major impact on milk gel properties (Chandrapala et al., 2013; Riener, Noci, Cronin, Morgan, & Lyng, 2009, 2010; Vercet, Oria, Marquina, Crelier, & Lopez-Buesa, 2002). In yogurt production, one of the main challenges is satisfactory, shown through characteristic textural properties in terms of soft and viscous product with a creamy texture, reduction of whey and slightly acidic taste. Basic steps in yogurt production gather acidification, fermentation and heat treatment. During fermentation, a textural defect of yogurt called graininess, which has a major impact on the sensorial properties of the final product, especially visual assessment, and in-mouth perception, often occurs. The main cause of graininess occurrence is particles of larger dimension, which besides visual appearance also affect the rheological properties, viscosity, and firmness (Sonne, Busch-Stockfisch, Weiss, & Hinrichs, 2014; van Marle, van den Ende, de Kruif, & Mellema, 1999). Nöbel, Protte, Körzendörfer, Hitzmann, and Hinrichs (2016) produced 26 stirred samples of yogurt and sonicated them with a frequency of 35 kHz for 5 min. The authors concluded that sonication resulted in the apparition of larger particles, while the increase of dry matter content affected the rheological properties of yogurt. Two important processes in the dairy industry, where ultrasound finds a significant implementation, are emulsification and homogenization (Brnčić et al., 2008). In both processes, ultrasound shows several crucial advantages that are reflected in significant energy savings (energy efficiency), higher emulsion stability, a significant reduction of surfactants uses as well as controllable size distributions. All mentioned advantages give many benefits of using ultrasound over conventional methods such as mechanical shaking, high- or ultra-high-pressure homogenizing and microfluidizing (Abismaïl, Canselier, Wilhelm, Delmas, & Gourdon, 1999; Juang & Lin, 2004). Cavitation that develops during sonication causes bubble collapse near or at the oil-water interface, resulting in disruption and mixing of two phases forming a fine emulsion (Thompson & Doraiswamy, 1999). One of the application possibilities of ultrasonic emulsification was demonstrated by Shanmugam and Ashokkumar (2014) through the incorporation of food oils in milk systems. The authors used ultrasound for the incorporation of flaxseed oil in skim milk. For emulsification of flaxseed oil in skim milk, 20 kHz system with an ultrasonic probe was used without any need for surfactants. Treated samples showed sufficient stability for even 9 days. Similar to the described ones, other scientific studies proved successful sonication applications of emulsification in milk systems with improved stability (Jafari, Assadpoor, He, & Bhandari, 2008; Jafari, He, & Bhandari, 2007; Jincai, Shaoying, & Rixian, 2013; Lad & Murthy, 2012; O'Sullivan, Murray, Flynn, & Norton, 2016; Yanjun et al., 2014). Some dairy products (e.g., milk, yogurt, ice cream, etc.) require first for their production, a homogenization step, allowing the improvement of stability against creaming during storage. Ultrasound technology finds its application in homogenization with similar advantages such as in emulsification. Number of scientific researches proved exceptional efficiency of sonication in homogenization of milk as well as on some physical properties of milk that is homogenized (Behrend & Schubert, 2001; Bosiljkov et al., 2011; Ertugay, Sengül, & Sengül, 2004; Koh et al., 2014; Sfakianakis, Topakas, & Tzia, 2015; Villamiel & de Jong, 2000; Wu, Hulbert, & Mount, 2000). Besides mentioned, the crystallization process is of great importance in the dairy industry, primarily for some procedures such as lactose removal. Except for the conventional methods in which the main disadvantages are long induction time and slow crystallization rate, ultrasound technology found a significant possibility to control crystallization in a process commonly known as sonocrystallization. During this process, crystal structure (shape and rate of crystallization) is controlled, crystallization process is significantly faster with higher efficacy and the whole process leads to cost-effectiveness in regards to conventional methods (Deora et al., 2013; Frydenberg, Hammershøj, Andersen, & Wiking, 2013). In some research studies, a rapid recovery of lactose by ultrasound-assisted crystallization was proved (Bund and Pandit, 2007a, 2007b; Patel and Murthy, 2009, 2011). For example, Patel and Murthy obtained lactose recovery yields in the range of 80–92% within 4 min of sonication (Patel & Murthy, 2009, 2011). Similarly, Zamanipoor, Dincer, Zisu, and Jayasena (2013) even found 5.6 times higher lactose yields in a system with an ultrasonic probe at a frequency of 20 kHz, without using solvents. One of the crucial benefits of sonocrystallization is also a significant reduction of induction times, proved by Dincer et al. (2014), where sonication decreased the induction time up to 3 min. Moreover, sonication shows considerable influence in the process of fat crystallization, which ultimately reflects on the structure and texture of product with the main claim to size and shape of crystals formed during crystallization (Hartel, 2013; Suzuki, Lee, Padilla, & Martini, 2010). Martini, Suzuki, and Hartel (2008) showed an application of ultrasound in crystallization of anhydrous milk fat that ultrasound reduces the induction time and simultaneously generates smaller crystals and higher viscosities. The process of fat crystallization is of great importance in the production of chocolate, especially in order to avoid undesirable fat bloom of chocolate, which may occur during the first cooling step of traditional tempering as a result of crystallized unstable polymorphic forms of cocoa butter (Afoakwa, Paterson, & Fowler, 2007). In order to obtain a stable product, Higaki, Ueno, Koyano, and Sato (2001) conducted sonocrystallization of cocoa butter at a frequency of 20 kHz in the range of 100–300 W powers for 3 s and concluded that the stable form of butter is directly crystallized without the formation of subsequent unstable forms, which leads to the conclusion that ultrasound is effective in controlling polymorphic crystallization of fats.

3.2.3 Sensorial Properties

3.2.3.1 Fruits and Vegetables

The sensorial quality of fruits, vegetables, and their products is based on several properties of which the most important are: texture, color, and flavor. During technological processing, the mentioned properties are strongly influenced by some factors (e.g., pH, oxygen presence, temperature, light, etc.). Non-invasive processing techniques have a positive effect on the preservation of the organoleptic properties of a product. One of the most common problems in the storage and processing of fruits and vegetables is browning, caused by the activity of endogenous enzyme PPO (polyphenoloxidase). Ultrasound is effective in the inactivation of enzymes, related to the color changes, but also in inactivating microorganisms, which cause food spoilage. Sulaiman et al. (2017) studied the sensory attributes of apple juice processed by thermosonication compared to the thermal processing and concluded that thermosonicated apple juices are long-stable, PPO inactivation was satisfactory and also the overall taste of such treated juices was acceptable for consumers. In sonication of grapefruit juice, in research conducted by Aadil et al. (2015), color values, cloud value and non-enzymatic browning (NEB) were evaluated as the main organoleptic characteristics. Cloud stability is an important quality parameter, which plays a significant role in color and flavor of fruit juices. Cloud value is related to the amount of particles composed of proteins, pectin, cellulose, etc. Cloud values of sonication grapefruit juices increased after processing as a result of the breakdown of bigger molecules into smaller due to the cavitation process. In the same research, the analyzed color values ($L*$ (whiteness), $a*$ (redness) and $b*$ (yellowness)) decreased during sonication. Related to the color change, NEB values increased during sonication. Similar results were reported by Tiwari,

Muthukumarappan, O'Donnell, and Cullen (2008) for orange juice samples. Color degradation (L∗, a∗, b∗ values decreased) was also observed during sonication of tomato juice (Adekunte et al., 2010). Šimunek et al. (2013) analyzed the aroma profile and sensory parameters (organoleptic assaying) of sonicated apple juice and revealed that sonication significantly influences the aroma profile, sensory properties, and color parameters from which the best results from sensory evaluation had ultrasonically-treated juice. Furthermore, many other research studies cited significant color changes of sonicated fruit and vegetable samples that can be attributed to the oxidation reactions occurring as a result of free radicals interaction, induced by cavitation (Aadil et al., 2013; Abid et al., 2013; Alexandre et al., 2012; Bhat et al., 2011; Dias et al., 2015; Pérez-Grijalba et al., 2017; Zou et al., 2016, 2017). For the mentioned reason, color stability of different fruit and vegetable samples during sonication can be significantly affected by key ultrasound factors: amplitude (power) and time. Its optimization is crucial to obtain a stable product of acceptable organoleptic properties, microbiological safety, and nutritional quality.

3.2.3.2 Meat Products

In recent times, scientific studies are still being conducted for the improvement of meat quality, especially texture and sensorial properties. Sonication strongly affects the meat quality properties, considering that thermal stability and texture attributes of collagen are directly associated with its influence on the connective meat tissues. Chang, Xu, Zhou, Li, and Huang (2012) treated beef semitendinosus muscle samples with high power ultrasound at the frequency of 40 kHz, a power of 1500 W, and for 10, 20, 30, 40, 50, and 60 min in an ultrasonic bath in order to investigate the effect of high-power ultrasound on the thermal characteristics of collagen and to evaluate the characteristics of collagen changes on meat quality (color) and texture properties. The main results from the presented study suggested that ultrasound did not significantly influence the L∗ and a∗ chromaticity values but significantly decreased the b∗ value when applied for 30 min. Changes were observed for fiber diameter and filtering residues, which values were reduced by ultrasound use. A significant decrease in enzyme activity (β-galactosidase and β-glucuronidase) was indicated when muscle samples were sonicated for 10 min. In addition, the authors of the conducted experiment confirmed sensitivity and decrease of collagen stability in sonicated samples. By Scanning Electron Microscopy (SEM) technique, denaturing, granulation, and aggregation of collagen fiber was observed in extracellular space. Cichoski et al. (2015) evaluated chromaticity parameters (overall color) of sonicated hot dog sausages. The results showed no significant differences of measured L, a∗ and b∗ values for inner and outer parts of sausages. However, slight decreases of a∗ and b∗ parameters were observed during storage in all treatments (conventional and ultrasound). Besides color evaluation, the authors evaluated the effect of ultrasound on some textural properties: hardness (N), elasticity (cm), cohesiveness and chewiness (N/cm) during which no significant changes were observed for samples treated by ultrasound. Similar results were obtained by Dolatowski and Stadnik (2007) who concluded that ultrasound treatment did not affect the meat lightness.

3.2.3.3 Dairy Products

It is important to emphasize that during sonocrystallization of systems with fats and oils, oxidation occurs forming thus free radicals which may cause further negative effects (Riener et al., 2009). Besides the formation of free radicals, oxidation process in systems with fats and oils often cause the formation of undesirable off-flavors, which formation is avoided by using ultrasound (Patrick, Blindt, & Janssen, 2004). Increased ultrasound intensity and duration of the treatment are key factors affecting the occurrence of undesirable burnt off-flavor, therefore the optimization of the mentioned factors is crucial to avoid adverse sensorial properties (Marchesini et al., 2012). Juliano et al. (2014) studied the effect of sonication on lipid oxidation in various types of milk at different ultrasound frequencies (20, 400, 1000, 1600 and 2000 kHz), using different temperatures (4, 20, 45, 63 °C), sonication times, and ultrasound energy inputs. Results of the study showed that lipid oxidation in milk can be optimized by controlling the ultrasound processing factors, including the frequency, the power levels, the processing time, the temperature of the system, and the fat content. Another dairy product, which during production assumes crystallization and in which ultrasound found desirable application, is ice cream. One of the main quality characteristics of ice cream is texture and taste directly influenced by ice crystal size. Mortazavi and Tabatabai (2008) managed to obtain sonicated ice cream with flavor and texture better than the control sample. More literature data also present a positive influence of sonication on sensorial and textural qualities of ice cream (Chow, Blindt, Chivers, & Povey, 2003; Mortazavi & Tabatabai, 2008).

3.3 Conclusions

There are many reasons for the comprehensive usage of ultrasound as an innovative technology in the processing of fruits and vegetables, meat and dairy products. Widely recognized as non-thermal and environmentally friendly, ultrasound has many benefits such as improved yield, better emulsification, shortened treatment time, lower energy consumption and less or no waste generation. It can influence more acceptable textural and sensory properties on vegetables and fruits as attributes but also can result in improved tenderization of meat products. Furthermore, ultrasound could in both batch and continuous modes lead towards more uniform homogenization of milk with smaller fat globules. Easy to handle and acceptable as a primary investment, ultrasound is not only able to be introduced in SME-s and large industry as separate technology, but also to improve existing technologies.

References

Aadil, R. M., Zeng, X.-A., Ali, A., Zeng, F., Farooq, M. A., Han, Z., … Jabbar, S. (2015). Influence of different pulsed electric field strengths on the quality of the grapefruit juice. *International Journal of Food Science and Technology, 50*(10), 2290–2296.

Aadil, R. M., Zeng, X.-A., Han, Z., & Sun, D.-W. (2013). Effects of ultrasound treatments on quality of grapefruit juice. *Food Chemistry, 141*(3), 3201–3206.

Abid, M., Jabbar, S., Wu, T., Hashim, M. M., Hu, B., Lei, S., … Zeng, X. (2013). Effect of ultrasound on different quality parameters of apple juice. *Ultrasonics Sonochemistry, 20*(5), 1182–1187.

Abismaïl, B., Canselier, J. P., Wilhelm, A. M., Delmas, H., & Gourdon, C. (1999). Emulsification by ultrasound: Drop size distribution and stability. *Ultrasonics Sonochemistry, 6*(1), 75–83.

Aburto, N. J., Ziolkovska, A., Hooper, L., Elliott, P., Cappuccio, F. P., & Meerpohl, J. J. (2013). Effect of lower sodium intake on health: Systematic review and meta-analyses. *BMJ, 346*, f1326.

Adekunte, A. O., Tiwari, B. K., Cullen, P. J., Scannell, A. G. M., & O'Donnell, C. P. (2010). Effect of sonication on colour, ascorbic acid and yeast inactivation in tomato juice. *Food Chemistry, 122*(3), 500–507.

Afoakwa, E. O., Paterson, A., & Fowler, M. (2007). Factors influencing rheological and textural qualities in chocolate – a review. *Trends in Food Science & Technology, 18*(6), 290–298.

Alarcon-Rojo, A. D., Janacua, H., Rodriguez, J. C., Paniwnyk, L., & Mason, T. J. (2015). Power ultrasound in meat processing. *Meat Science, 107*, 86–93.

Alexandre, E. M. C., Brandão, T. R. S., & Silva, C. L. M. (2012). Efficacy of non-thermal technologies and sanitizer solutions on microbial load reduction and quality retention of strawberries. *Journal of Food Engineering, 108*(3), 417–426.

Alighourchi, H. R., Barzegar, M., Sahari, M. A., & Abbasi, S. (2013). Effect of sonication on anthocyanins, total phenolic content, and antioxidant capacity of pomegranate juices. *International Food Research Journal, 20*(4), 1703–1709.

Alligar, H. (1975). Ultrasonic disruption. *American Laboratory, 10*, 75–85.

Alvarado, C., & McKee, S. (2007). Marination to improve functional properties and safety of poultry meat. *The Journal of Applied Poultry Research, 16*(1), 113–120.

Arzeni, C., Martínez, K., Zema, P., Arias, A., Pérez, O. E., & Pilosof, A. M. R. (2012). Comparative study of high intensity ultrasound effects on food proteins functionality. *Journal of Food Engineering, 108*(3), 463–472.

Ashokkumar, M., Kentish, S., Lee, J., Zisu, B., Palmer, M., & Augustin, M. (2009). *Processing of dairy ingredients by ultrasonication*. Patent number: WO2009/079691A1.

Ashokkumar, M., Lee, J., Zisu, B., Bhaskarcharya, R., Palmer, M., & Kentish, S. (2009). Hot topic: Sonication increases the heat stability of whey proteins. *Journal of Dairy Science, 92*(11), 5353–5356.

Ashokkumar, M., Sunartio, D., Kentish, S., Mawson, R., Simons, L., Vilkhu, K., & Versteeg, C. K. (2008). Modification of food ingredients by ultrasound to improve functionality: A preliminary study on a model system. *Innovative Food Science & Emerging Technologies, 9*(2), 155–160.

Barba, F. J., Zhu, Z., Koubaa, M., Sant'Ana, A. S., & Orlien, V. (2016). Green alternative methods for the extraction of antioxidant bioactive compounds from winery wastes and by-products: A review. *Trends in Food Science & Technology, 49*, 96–109.

Barekat, S., & Soltanizadeh, N. (2017). Improvement of meat tenderness by simultaneous application of high-intensity ultrasonic radiation and papain treatment. *Innovative Food Science & Emerging Technologies, 39*, 223–229.

Behrend, O., & Schubert, H. (2001). Influence of hydrostatic pressure and gas content on continuous ultrasound emulsification. *Ultrasonics Sonochemistry, 8*(3), 271–276.

Bernal, V., & Jelen, P. (1985). Thermal stability of whey proteins – a calorimetric study. *Journal of Dairy Science, 68*(11), 2847–2852.

Bhat, R., Kamaruddin, N. S. B. C., Min-Tze, L., & Karim, A. A. (2011). Sonication improves kasturi lime (*Citrus microcarpa*) juice quality. *Ultrasonics Sonochemistry, 18*(6), 1295–1300.

Bosiljkov, T., Tripalo, B., Brnčić, M., Ježek, D., Karlović, S., & Jagušt, I. (2011). Influence of high intensity ultrasound with different probe diameter on the degree of homogenization (variance) and physical properties of cow milk. *African Journal of Biotechnology, 10*(1), 34–41.

Brnčić, M., Bosiljkov, T., Ukrainczyk, M., Tripalo, B., Brnčić, S. R., Karlović, S., ... Topić, D. V. (2011). Influence of whey protein addition and feed moisture content on chosen physicochemical properties of directly expanded corn extrudates. *Food and Bioprocess Technology, 4*(7), 1296–1306.

Brnčić, M., Ježek, D., Rimac Brnčić, S., Bosiljkov, T., & Tripalo, B. (2008). Influence of whey protein concentrate addition on textural properties of corn flour extrudates. *Mljekarstvo, 58*(2), 131–149.

Brnčić, M., Karlović, S., Rimac Brnčić, S., Bosiljkov, T., Ježek, D., & Tripalo, B. (2010). Textural properties of infra red dried apple slices as affected by high power ultrasound pre-treatment. *African Journal of Biotechnology, 9*(41), 6907–6915.

Brnčić, M., Tripalo, B., Brncic, S. R., Karlovic, S., Zupan, A., & Herceg, Z. (2009). Evaluation of textural properties for whey enriched direct extruded and puffed corn based products. *Bulgarian Journal of Agricultural Science, 15*(3), 204–213.

Brnčić, M., Tripalo, B., Penava, A., Karlović, D., Ježek, D., Vikić Topić, D., ... Bosiljov, T. (2009). Applications of power ultrasound for foodstuffs processing. *Croatian Journal of Food Technology, Biotechnology and Nutrition, 1–2*, 32–37.

Bund, R. K., & Pandit, A. B. (2007a). Rapid lactose recovery from paneer whey using sonocrystallization: A process optimization. *Chemical Engineering and Processing: Process Intensification, 46*(9), 846–850.

Bund, R. K., & Pandit, A. B. (2007b). Sonocrystallization: Effect on lactose recovery and crystal habit. *Ultrasonics Sonochemistry, 14*(2), 143–152.

Cao, S., Hu, Z., Pang, B., Wang, H., Xie, H., & Wu, F. (2010). Effect of ultrasound treatment on fruit decay and quality maintenance in strawberry after harvest. *Food Control, 21*(4), 529–532.

Carbonell-Capella, J. M., Barba, F. J., Esteve, M. J., & Frígola, A. (2014). Quality parameters, bioactive compounds and their correlation with antioxidant capacity of commercial fruit-based baby foods. *Food Science and Technology International, 20*(7), 479–487.

Cárcel, J. A., Benedito, J., Bon, J., & Mulet, A. (2007). High intensity ultrasound effects on meat brining. *Meat Science, 76*(4), 611–619.

Casiraghi, E., Alamprese, C., & Pompei, C. (2007). Cooked ham classification on the basis of brine injection level and pork breeding country. *LWT – Food Science and Technology, 40*(1), 164–169.

Chandrapala, J., Martin, G. J. O., Zisu, B., Kentish, S. E., & Ashokkumar, M. (2012). The effect of ultrasound on casein micelle integrity. *Journal of Dairy Science, 95*(12), 6882–6890.

Chandrapala, J., Oliver, C., Kentish, S., & Ashokkumar, M. (2012). Ultrasonics in food processing. *Ultrasonics Sonochemistry, 19*(5), 975–983.

Chandrapala, J., Zisu, B., Kentish, S., & Ashokkumar, M. (2012). The effects of high-intensity ultrasound on the structural and functional properties of α-Lactalbumin, β-Lactoglobulin and their mixtures. *Food Research International, 48*(2), 940–943.

Chandrapala, J., Zisu, B., Kentish, S., & Ashokkumar, M. (2013). Influence of ultrasound on chemically induced gelation of micellar casein systems. *The Journal of Dairy Research, 80*(2), 138–143.

Chandrapala, J., Zisu, B., Palmer, M., Kentish, S., & Ashokkumar, M. (2011). Effects of ultrasound on the thermal and structural characteristics of proteins in reconstituted whey protein concentrate. *Ultrasonics Sonochemistry, 18*(5), 951–957.

Chang, H.-J., Xu, X.-L., Zhou, G.-H., Li, C.-B., & Huang, M. (2012). Effects of characteristics changes of collagen on meat physicochemical properties of beef semitendinosus muscle during ultrasonic processing. *Food and Bioprocess Technology, 5*(1), 285–297.

Chemat, F., Zill-e-Huma, & Khan, M. K. (2011). Applications of ultrasound in food technology: Processing, preservation and extraction. *Ultrasonics Sonochemistry, 18*(4), 813–835.

Chow, R., Blindt, R., Chivers, R., & Povey, M. (2003). The sonocrystallisation of ice in sucrose solutions: Primary and secondary nucleation. *Ultrasonics, 41*(8), 595–604.

Cichoski, A. J., Rampelotto, C., Silva, M. S., de Moura, H. C., Terra, N. N., Wagner, R., … Barin, J. S. (2015). Ultrasound-assisted post-packaging pasteurization of sausages. *Innovative Food Science & Emerging Technologies, 30*, 132–137.

Coakley, W. T., & Nyborg, W. L. (1978). Ultrasound: Its applications in medicine and biology, part 1. In *Cavitation; dynamics of gas bubbles; applications* (pp. 77–153). New York, NY: Elsevier Scientific Publishing.

Deora, N. S., Misra, N. N., Deswal, A., Mishra, H. N., Cullen, P. J., & Tiwari, B. K. (2013). Ultrasound improved crystallization in food processing. *Food Engineering Reviews, 5*(1), 36–44.

Dias, D. d. R. C., Barros, Z. M. P., de Carvalho, C. B. O., Honorato, F. A., Guerra, N. B., & Azoubel, P. M. (2015). Effect of sonication on soursop juice quality. *LWT – Food Science and Technology, 62*(1, Part 2), 883–889.

Dickens, J. A., Lyon, C. E., & Wilson, R. L. (1991). Effect of ultrasonic radiation on some physical characteristics of broiler breast muscle and cooked meat. *Poultry Science, 70*(2), 389–396.

Dincer, T. D., Zisu, B., Vallet, C. G. M. R., Jayasena, V., Palmer, M., & Weeks, M. (2014). Sonocrystallisation of lactose in an aqueous system. *International Dairy Journal, 35*(1), 43–48.

Dolatowski, Z. J., & Stadnik, J. (2007). Effect of sonication on technological properties of beef. *Visnyk of Dnipropetrovsk University. Biology, Ecology, 15*(1), 220–223.

Drosou, C., Kyriakopoulou, K., Bimpilas, A., Tsimogiannis, D., & Krokida, M. (2015). A comparative study on different extraction techniques to recover red grape pomace polyphenols from vinification byproducts. *Industrial Crops and Products, 75*, 141–149.

Dujmić, F., Brnčić, M., Karlović, S., Bosiljkov, T., Ježek, D., Tripalo, B., & Mofardin, I. (2013). Ultrasound-assisted infrared drying of pear slices: Textural issues. *Journal of Food Process Engineering, 36*(3), 397–406.

Ertugay, M. F., Sengül, M., & Sengül, M. (2004). Effect of ultrasound treatment on milk homogenization and particle size distribution of fat. *Turkish Journal of Veterinary & Animal Sciences, 28*, 303–308.

Fernandes, F. A. N., & Rodrigues, S. (2007). Ultrasound as pre-treatment for drying of fruits: Dehydration of banana. *Journal of Food Engineering, 82*(2), 261–267.

Frydenberg, R. P., Hammershøj, M., Andersen, U., & Wiking, L. (2013). Ultrasonication affects crystallization mechanisms and kinetics of anhydrous milk fat. *Crystal Growth & Design, 13*(12), 5375–5382.

Galvan d'Alessandro, L., Kriaa, K., Nikov, I., & Dimitrov, K. (2012). Ultrasound assisted extraction of polyphenols from black chokeberry. *Separation and Purification Technology, 93*, 42–47.

González-González, L., Luna-Rodríguez, L., Carrillo-López, L. M., Alarcón-Rojo, A. D., García-Galicia, I., & Reyes-Villagrana, R. (2017). Ultrasound as an alternative to conventional marination: Acceptability and mass transfer. *Journal of Food Quality*. https://doi.org/10.1155/2017/8675720

Hartel, R. W. (2013). Advances in food crystallization. *Annual Review of Food Science and Technology, 4*, 277–292.

Higaki, K., Ueno, S., Koyano, T., & Sato, K. (2001). Effects of ultrasonic irradiation on crystallization behavior of tripalmitoylglycerol and cocoa butter. *Journal of the American Oil Chemists' Society, 78*(5), 513–518.

Hu, H., Wu, J., Li-Chan, E. C. Y., Zhu, L., Zhang, F., Xu, X., … Pan, S. (2013). Effects of ultrasound on structural and physical properties of soy protein isolate (SPI) dispersions. *Food Hydrocolloids, 30*(2), 647–655.

Huff-Lonergan, E. (2010). Water-holding capacity of fresh meat. *American Meat Science Association*, 1–8.

Huff-Lonergan, E., & Lonergan, S. M. (2005). Mechanisms of water-holding capacity of meat: The role of postmortem biochemical and structural changes. *Meat Science, 71*(1), 194–204.
Inguglia, E. S., Zhang, Z., Tiwari, B. K., Kerry, J. P., & Burgess, C. M. (2017). Salt reduction strategies in processed meat products – A review. *Trends in Food Science & Technology, 59*, 70–78.
Istrati, D. (2008). The influence of enzymatic tenderization with papain on functional properties of adult beef. *Journal of Agroalimentary Processes and Technologies, 14*, 140–146.
Jafari, S. M., Assadpoor, E., He, Y., & Bhandari, B. (2008). Re-coalescence of emulsion droplets during high-energy emulsification. *Food Hydrocolloids, 22*(7), 1191–1202.
Jafari, S. M., He, Y., & Bhandari, B. (2007). Production of sub-micron emulsions by ultrasound and microfluidization techniques. *Journal of Food Engineering, 82*(4), 478–488.
Jiménez-Colmenero, F. (2013). Potential applications of multiple emulsions in the development of healthy and functional foods. *Food Research International, 52*(1), 64–74.
Jincai, Z., Shaoying, Z., & Rixian, Y. (2013). Ultrasonic enhanced walnut protein emulsifying property. *Journal of Food Processing & Technology, 4*(7). https://doi.org/10.4172/2157-7110.1000244
Juang, R.-S., & Lin, K.-H. (2004). Ultrasound-assisted production of W/O emulsions in liquid surfactant membrane processes. *Colloids and Surfaces A: Physicochemical and Engineering Aspects, 238*(1), 43–49.
Juliano, P., Torkamani, A. E., Leong, T., Kolb, V., Watkins, P., Ajlouni, S., & Singh, T. K. (2014). Lipid oxidation volatiles absent in milk after selected ultrasound processing. *Ultrasonics Sonochemistry, 21*(6), 2165–2175.
Kang, D.-C., Gao, X.-Q., Ge, Q.-F., Zhou, G.-H., & Zhang, W.-G. (2017). Effects of ultrasound on the beef structure and water distribution during curing through protein degradation and modification. *Ultrasonics Sonochemistry, 38*, 317–325.
Kiani, H., Sun, D.-W., & Zhang, Z. (2013). Effects of processing parameters on the convective heat transfer rate during ultrasound assisted low temperature immersion treatment of a stationary sphere. *Journal of Food Engineering, 115*(3), 384–390.
Knorr, D., Zenker, M., Heinz, V., & Lee, D.-U. (2004). Applications and potential of ultrasonics in food processing. *Trends in Food Science & Technology, 15*(5), 261–266.
Koh, L. L. A., Chandrapala, J., Zisu, B., Martin, G. J. O., Kentish, S. E., & Ashokkumar, M. (2014). A comparison of the effectiveness of sonication, high shear mixing and homogenization on improving the heat stability of whey proteins solutions. *Food and Bioprocess Technology, 7*(2), 556–566.
Kuijpers, M. W. A., Kemmere, M. F., & Keurentjes, J. T. F. (2002). Calorimetric study of the energy efficiency for ultrasound-induced radical formation. *Ultrasonics, 40*(1–8), 675–678.
Lad, V. N., & Murthy, Z. V. P. (2012). Enhancing the stability of oil-in-water emulsions emulsified by coconut milk protein with the application of acoustic cavitation. *Industrial & Engineering Chemistry Research, 51*(11), 4222–4229.
Lawrie, R., & Ledward, D. (2006). *Lawrie's meat science* (7th ed.). Cambridge, UK: Woodhead Publishing.
Lianfu, Z., & Zelong, L. (2008). Optimization and comparison of ultrasound/microwave assisted extraction (UMAE) and ultrasonic assisted extraction (UAE) of lycopene from tomatoes. *Ultrasonics Sonochemistry, 15*(5), 731–737.
Lim, S. S., Vos, T., Flaxman, A. D., Danaei, G., Shibuya, K., Adair-Rohani, H., … Memish, Z. A. (2012). A comparative risk assessment of burden of disease and injury attributable to 67 risk factors and risk factor clusters in 21 regions, 1990-2010: A systematic analysis for the Global Burden of Disease Study 2010. *Lancet, 380*(9859), 2224–2260.
Liu, Z., Juliano, P., Williams, R. P., Niere, J., & Augustin, M. A. (2014a). Ultrasound effects on the assembly of casein micelles in reconstituted skim milk. *Journal of Dairy Research, 81*(2), 146–155.
Liu, Z., Juliano, P., Williams, R. P., Niere, J., & Augustin, M. A. (2014b). Ultrasound improves the renneting properties of milk. *Ultrasonics Sonochemistry, 21*(6), 2131–2137.

Lyng, J. G., Allen, P., & McKenna, B. M. (1997). The influence of high intensity ultrasound bath on aspects of beef tenderness. *Journal of Muscle Foods, 8*(3), 237–249.

Lyng, J. G., Allen, P., & McKenna, B. M. (1998). The effect on aspects of beef tenderness of pre- and post-rigor exposure to a high intensity ultrasound probe. *Journal of the Science of Food and Agriculture, 78*(3), 308–314.

Madadlou, A., Mousavi, M. E., Emam-djomeh, Z., Ehsani, M., & Sheehan, D. (2009). Sonodisruption of re-assembled casein micelles at different pH values. *Ultrasonics Sonochemistry, 16*(5), 644–648.

Mane, S., Bremner, D. H., Tziboula-Clarke, A., & Lemos, M. A. (2015). Effect of ultrasound on the extraction of total anthocyanins from Purple Majesty potato. *Ultrasonics Sonochemistry, 27*, 509–514.

Mapiye, C., Vahmani, P., Mlambo, V., Muchenje, V., Dzama, K., Hoffman, L. C., & Dugan, M. E. R. (2015). The trans-octadecenoic fatty acid profile of beef: Implications for global food and nutrition security. *Food Research International, 76*, 992–1000.

Marchesini, G., Balzan, S., Montemurro, F., Fasolato, L., Andrighetto, I., Segato, S., & Novelli, E. (2012). Effect of ultrasound alone or ultrasound coupled with CO_2 on the chemical composition, cheese-making properties and sensory traits of raw milk. *Innovative Food Science and Emerging Technologies, 16*, 391–397.

Martini, S., Suzuki, A. H., & Hartel, R. W. (2008). Effect of high intensity ultrasound on crystallization behavior of anhydrous milk fat. *Journal of the American Oil Chemists' Society, 85*(7), 621–628.

Medina-Torres, N., Ayora-Talavera, T., Espinosa-Andrews, H., Sánchez-Contreras, A., & Pacheco, N. (2017). Ultrasound assisted extraction for the recovery of phenolic compounds from vegetable sources. *Agronomy, 7*(3), 47.

Mieszczakowska-Frąc, M., Dyki, B., & Konopacka, D. (2016). Effects of ultrasound on polyphenol retention in apples after the application of predrying treatments in liquid medium. *Food and Bioprocess Technology, 9*(3), 543–552.

Mortazavi, A., & Tabatabai, F. (2008). Study of ice cream freezing process after treatment with ultrasound. *World Applied Sciences Journal, 4*(2), 188–190.

Ngamwonglumlert, L., Devahastin, S., & Chiewchan, N. (2017). Natural colorants: Pigment stability and extraction yield enhancement via utilization of appropriate pretreatment and extraction methods. *Critical Reviews in Food Science and Nutrition, 57*(15), 3243–3259.

Nguyen, N. H. A., & Anema, S. G. (2010). Effect of ultrasonication on the properties of skim milk used in the formation of acid gels. *Innovative Food Science & Emerging Technologies, 11*(4), 616–622.

Nishihara, T., & Doty, P. (1958). The sonic fragmentation of collagen macromolecules. *Proceedings of the National Academy of Sciences of the United States of America, 44*(5), 411–417.

Nöbel, S., Protte, K., Körzendörfer, A., Hitzmann, B., & Hinrichs, J. (2016). Sonication induced particle formation in yogurt: Influence of the dry matter content on the physical properties. *Journal of Food Engineering, 191*, 77–87.

Nowacka, M., Wiktor, A., Śledź, M., Jurek, N., & Witrowa-Rajchert, D. (2012). Drying of ultrasound pretreated apple and its selected physical properties. *Journal of Food Engineering, 113*(3), 427–433.

Ojha, K. S., Perussello, C. A., García, C. Á., Kerry, J. P., Pando, D., & Tiwari, B. K. (2017). Ultrasonic-assisted incorporation of nano-encapsulated omega-3 fatty acids to enhance the fatty acid profile of pork meat. *Meat Science, 132*, 99–106.

Olmedilla-Alonso, B., Jiménez-Colmenero, F., & Sánchez-Muniz, F. J. (2013). Development and assessment of healthy properties of meat and meat products designed as functional foods. *Meat Science, 95*(4), 919–930.

Ordóñez-Santos, L. E., Martínez-Girón, J., & Arias-Jaramillo, M. E. (2017). Effect of ultrasound treatment on visual color, vitamin C, total phenols, and carotenoids content in Cape gooseberry juice. *Food Chemistry, 233*, 96–100.

O'Sullivan, J., Murray, B., Flynn, C., & Norton, I. (2016). The effect of ultrasound treatment on the structural, physical and emulsifying properties of animal and vegetable proteins. *Food Hydrocolloids, 53*, 141–154.

Ozuna, C., Puig, A., García-Pérez, J. V., Mulet, A., & Cárcel, J. A. (2013). Influence of high intensity ultrasound application on mass transport, microstructure and textural properties of pork meat (*Longissimus dorsi*) brined at different NaCl concentrations. *Journal of Food Engineering, 119*(1), 84–93.

Pan, Z., Qu, W., Ma, H., Atungulu, G. G., & McHugh, T. H. (2011). Continuous and pulsed ultrasound-assisted extractions of antioxidants from pomegranate peel. *Ultrasonics Sonochemistry, 18*(5), 1249–1257.

Patel, S. R., & Murthy, V. P. (2011). Waste valorisation: Recovery of lactose from partially deproteonated whey by using acetone as antisolvent. *Dairy Science Technology, 91*, 53–63.

Patel, S. R., & Murthy, Z. V. P. (2009). Ultrasound assisted crystallization for the recovery of lactose in an anti-solvent acetone. *Crystal Research and Technology, 44*(8), 889–896.

Patrick, M., Blindt, R., & Janssen, J. (2004). The effect of ultrasonic intensity on the crystal structure of palm oil. *Ultrasonics Sonochemistry, 11*(3), 251–255.

Paulsson, M., & Dejmek, P. (1990). Thermal denaturation of whey proteins in mixtures with caseins studied by differential scanning calorimetry. *Journal of Dairy Science, 73*(3), 590–600.

Pérez-Grijalba, B., García-Zebadúa, J. C., Ruíz-Pérez, V. M., Téllez-Medina, D. I., Guzmán-Gerónimo, R. I., & Mora-Escobedo, R. (2017). Biofunctionality, colorimetric coefficients and microbiological stability of blackberry (*Rubus fructicosus* var. Himalaya) juice under microwave/ultrasound processing. *Revista Mexicana de Ingeniería Química, 17*(1), 13–28.

Pétrier, C., Combet, E., & Mason, T. (2007). Oxygen-induced concurrent ultrasonic degradation of volatile and non-volatile aromatic compounds. *Ultrasonics Sonochemistry, 14*(2), 117–121.

Pohlman, F. W., Dikeman, M. E., & Kropf, D. H. (1997). Effects of high intensity ultrasound treatment, storage time and cooking method on shear, sensory, instrumental color and cooking properties of packaged and unpackaged beef pectoralis muscle. *Meat Science, 46*(1), 89–100.

Riener, J., Noci, F., Cronin, D. A., Morgan, D. J., & Lyng, J. G. (2009). The effect of thermosonication of milk on selected physicochemical and microstructural properties of yoghurt gels during fermentation. *Food Chemistry, 114*(3), 905–911.

Riener, J., Noci, F., Cronin, D. A., Morgan, D. J., & Lyng, J. G. (2010). A comparison of selected quality characteristics of yoghurts prepared from thermosonicated and conventionally heated milks. *Food Chemistry, 119*(3), 1108–1113.

Roberts, R. T. (1991). Sound for processing food. *Nutrition & Food Science, 91*(3), 17–18.

Roncalés, P., Ceña, P., Beltrán, J. A., & Jaime, I. (1993). Ultrasonication of lamb skeletal muscle fibres enhances postmortem proteolysis. *Zeitschrift Fur Lebensmittel-Untersuchung Und -Forschung, 196*(4), 339–342.

Rosu, A.-M., Nistor, D. I., Miron, N. D., Popa, M. I., & Cojocaru, R. M. (2017). Ultrasound-assisted extraction of photosynthetic pigments from dried dill (*Anethum graveolens*). *Food and Environment Safety Journal, 11*(4), 5–9.

Ruusunen, M., & Puolanne, E. (2005). Reducing sodium intake from meat products. *Meat Science, 70*(3), 531–541.

Sfakianakis, P., Topakas, E., & Tzia, C. (2015). Comparative study on high-intensity ultrasound and pressure milk homogenization: Effect on the kinetics of yogurt fermentation process. *Food and Bioprocess Technology, 8*(3), 548–557.

Shanmugam, A., & Ashokkumar, M. (2014). Ultrasonic preparation of stable flax seed oil emulsions in dairy systems – Physicochemical characterization. *Food Hydrocolloids, 39*, 151–162.

Shanmugam, A., Chandrapala, J., & Ashokkumar, M. (2012). The effect of ultrasound on the physical and functional properties of skim milk. *Innovative Food Science & Emerging Technologies, 16*, 251–258.

Šic Žlabur, J., Dobričević, N., Pliestić, S., Galić, A., Bilić, D. P., & Voća, S. (2017). Antioxidant potential of fruit juice with added chokeberry powder (*Aronia melanocarpa*). *Molecules, 22*(12), 2158.

Šic Žlabur, J., Voća, S., Dobričević, N., Brnčić, M., Dujmić, F., & Rimac Brnčić, S. (2015). Optimization of ultrasound assisted extraction of functional ingredients from *Stevia rebaudiana* bertoni leaves. *International Agrophysics, 29*(2), 231–237.

Šimunek, M., Režek Jambrak, A., Petrović, M., Juretić, H., Major, N., Herceg, Z., … Vukušić, T. (2013). Aroma profile and sensory properties of ultrasound-treated apple juice and nectar. *Food Technology and Biotechnology, 51*(1), 101–111.

Siró, I., Vén, C., Balla, C., Jónás, G., Zeke, I., & Friedrich, L. (2009). Application of an ultrasonic assisted curing technique for improving the diffusion of sodium chloride in porcine meat. *Journal of Food Engineering, 91*(2), 353–362.

Smith, N. B., Cannon, J. E., Novakofski, J. E., McKeith, F. K., & O'Brien, W. D. (1991). Tenderization of semitendinosus muscle using high intensity ultrasound. In *IEEE 1991 Ultrasonics Symposium* (Vol. 2, pp. 1371–1374).

Sonne, A., Busch-Stockfisch, M., Weiss, J., & Hinrichs, J. (2014). Improved mapping of in-mouth creaminess of semi-solid dairy products by combining rheology, particle size, and tribology data. *LWT – Food Science and Technology, 59*(1), 342–347.

Soria, A. C., & Villamiel, M. (2010). Effect of ultrasound on the technological properties and bioactivity of food: A review. *Trends in Food Science & Technology, 21*(7), 323–331.

Stadnik, J., Dolatowski, Z. J., & Baranowska, H. M. (2008). Effect of ultrasound treatment on water holding properties and microstructure of beef (*m. semimembranosus*) during ageing. *LWT – Food Science and Technology, 41*(10), 2151–2158.

Stagni, N., & De Bernard, B. (1968). Lysosomal enzyme activity in rat and beef skeletal muscle. *Biochimica et Biophysica Acta, 170*, 129–139.

Suárez-Jacobo, A., Rüfer, C. E., Gervilla, R., Guamis, B., Roig-Sagués, A. X., & Saldo, J. (2011). Influence of ultra-high pressure homogenisation on antioxidant capacity, polyphenol and vitamin content of clear apple juice. *Food Chemistry, 127*(2), 447–454.

Sulaiman, A., Farid, M., & Silva, F. V. (2017). Quality stability and sensory attributes of apple juice processed by thermosonication, pulsed electric field and thermal processing. *Food Science and Technology International, 23*(3), 265–276.

Suzuki, A. H., Lee, J., Padilla, S. G., & Martini, S. (2010). Altering functional properties of fats using power ultrasound. *Journal of Food Science, 75*(4), E208–E214.

Tao, Y., García, J. F., & Sun, D.-W. (2014). Advances in wine aging technologies for enhancing wine quality and accelerating wine aging process. *Critical Reviews in Food Science and Nutrition, 54*(6), 817–835.

Tao, Y., & Sun, D.-W. (2015). Enhancement of food processes by ultrasound: A review. *Critical Reviews in Food Science and Nutrition, 55*(4), 570–594.

Tapasco, Z. Y., Restrepo, M. D. A., & Suarez, M. H. (2011). Efecto reologico de hidrocoloides sobre la salmuera de marinado de carne bovina. *Biotecnologıa En El Sector Agropecuario y Agroindustrial, 9*(2), 23–31.

Thompson, L. H., & Doraiswamy, L. K. (1999). Sonochemistry: Science and engineering. *Industrial & Engineering Chemistry Research, 38*(4), 1215–1249.

Tiwari, B. K., Muthukumarappan, K., O'Donnell, C. P., & Cullen, P. J. (2008). Colour degradation and quality parameters of sonicated orange juice using response surface methodology. *LWT – Food Science and Technology, 41*(10), 1876–1883.

Tiwari, B. K., O'Donnell, C. P., & Cullen, P. J. (2009). Effect of sonication on retention of anthocyanins in blackberry juice. *Journal of Food Engineering, 93*(2), 166–171.

Tiwari, B. K., O'Donnell, C. P., Muthukumarappan, K., & Cullen, P. J. (2008). Effect of ultrasound processing on the quality and nutritional properties of fruit juices. *Stewart Postharvest Review, 4*, 1–6.

Tiwari, B. K., O'Donnell, C. P., Patras, A., & Cullen, P. J. (2008). Anthocyanin and ascorbic acid degradation in sonicated strawberry juice. *Journal of Agricultural and Food Chemistry, 56*(21), 10071–10077.

Tiwari, B. K., Patras, A., Brunton, N., Cullen, P. J., & O'Donnell, C. P. (2010). Effect of ultrasound processing on anthocyanins and color of red grape juice. *Ultrasonics Sonochemistry, 17*(3), 598–604.

Tomšik, A., Pavlić, B., Vladić, J., Ramić, M., Brindza, J., & Vidović, S. (2016). Optimization of ultrasound-assisted extraction of bioactive compounds from wild garlic (*Allium ursinum* L.). *Ultrasonics Sonochemistry, 29*, 502–511.

Troy, D. J., Tiwari, B. K., & Joo, S.-T. (2016). Health implications of beef intramuscular fat consumption. *Korean Journal for Food Science of Animal Resources, 36*(5), 577–582.

Turantaş, F., Kılıç, G. B., & Kılıç, B. (2015). Ultrasound in the meat industry: General applications and decontamination efficiency. *International Journal of Food Microbiology, 198*, 59–69.

Uribe, E., Delgadillo, A., Giovagnoli-Vicuña, C., Quispe-Fuentes, I., & Zura-Bravo, L. (2015). Extraction techniques for bioactive compounds and antioxidant capacity determination of Chilean papaya (*Vasconcellea pubescens*) fruit. *Journal of Chemistry*. https://doi.org/10.1155/2015/347532

van Marle, M. E., van den Ende, D., de Kruif, C. G., & Mellema, J. (1999). Steady-shear viscosity of stirred yogurts with varying ropiness. *Journal of Rheology, 43*(6), 1643–1662.

Vercet, A., Oria, R., Marquina, P., Crelier, S., & Lopez-Buesa, P. (2002). Rheological properties of yoghurt made with milk submitted to manothermosonication. *Journal of Agricultural and Food Chemistry, 50*(21), 6165–6171.

Villamiel, M., & de Jong, P. (2000). Influence of high-intensity ultrasound and heat treatment in continuous flow on fat, proteins, and native enzymes of milk. *Journal of Agricultural and Food Chemistry, 48*(2), 472–478.

Wang, Z., Xu, W., Kang, N., Shen, Q., & Zhang, D. (2016). Microstructural, protein denaturation and water holding properties of lamb under pulse vacuum brining. *Meat Science, 113*, 132–138.

Wu, H., Hulbert, G. J., & Mount, J. R. (2000). Effects of ultrasound on milk homogenization and fermentation with yogurt starter. *Innovative Food Science & Emerging Technologies, 1*(3), 211–218.

Xargayó, M., Lagares, J. E., Fernández, D., Borrell, D., & Juncá, G. (2004). Solution for Improving meat texture. Influence of spray injection on the organoleptic and sensory characteristics. *Fleis-Chwirtschaft International, 2*, 68–74.

Yanjun, S., Jianhang, C., Shuwen, Z., Hongjuan, L., Jing, L., Lu, L., … Jiaping, L. (2014). Effect of power ultrasound pre-treatment on the physical and functional properties of reconstituted milk protein concentrate. *Journal of Food Engineering, 124*, 11–18.

Zamanipoor, M., Dincer, T., Zisu, B., & Jayasena, V. (2013). Nucleation and growth rates of lactose as affected by ultrasound in aqueous solutions. *Dairy Science and Technology, 93*, 595–604.

Zayas, J. F., & Strokova, N. D. (1972). Influence of ultrasound on properties of meat proteins. In *XVIII Eur. Congr. Meat Res. Workers* (pp. 206–213).

Zayatas, Y. (1971). Effect of ultrasound on animal tissues. *Myasnaya-Industriya SSSR, 42*(3), 33–35.

Zinoviadou, K. G., Galanakis, C. M., Brnčić, M., Grimi, N., Boussetta, N., Mota, M. J., … Barba, F. J. (2015). Fruit juice sonication: Implications on food safety and physicochemical and nutritional properties. *Food Research International, 77*(Part 4), 743–752.

Zisu, B., Bhaskaracharya, R., Kentish, S., & Ashokkumar, M. (2010). Ultrasonic processing of dairy systems in large scale reactors. *Ultrasonics Sonochemistry, 17*(6), 1075–1081.

Zou, Y., Hou, X., Zou, Y., & Hou, X. (2017). Sonication enhances quality and antioxidant activity of blueberry juice. *Food Science and Technology, 37*(4), 599–603.

Zou, Y., Jiang, A., Zou, Y., & Jiang, A. (2016). Effect of ultrasound treatment on quality and microbial load of carrot juice. *Food Science and Technology, 36*(1), 111–115.

Chapter 4
Impact of High-Pressure Processing on Food Quality

Mauro D. Santos, Rita S. Inácio, Liliana G. Fidalgo, Rui P. Queirós, Silvia A. Moreira, Ricardo V. Duarte, Ana M. P. Gomes, Ivonne Delgadillo, and Jorge A. Saraiva

4.1 Introduction

High-Pressure Processing (HPP) application is increasing at food industries as a possible alternative to heat treatments for food preservation and processing. It is being mainly applied to inactivate microorganisms and enzymes, with lower degradation of flavors and nutrients, minimizing the losses of beneficial ingredients, resulting in distinctive organoleptic properties of foods (Huppertz, Kelly, & Fox, 2002; Pasha, Saeed, Sultan, Khan, & Rohi, 2014). Since HPP acts on volume compression, due to the low change in volume on low-molecular compounds, such as vitamins and other functional compounds, the effects of this technology are expected to be minimum on these compounds unlike thermal treatment (Wang, Huang, Hsu, & Yang, 2016). Alike, HPP has also a lower effect on flavor and color compounds of food products, compared to the color changes and formation of off-flavors caused by thermal pasteurization (Wang, Huang, et al. 2016).

As the overall nutritional properties of foods can be better preserved by HPP, food texture can also be better maintained by this technology, affecting to a lesser extent the quality and acceptance of the products by the consumer, unlike thermal treatments that are more prone to affect texture and structure (i.e., due to loss of instrumental firmness by membrane disruption on vegetable/fruit products).

M. D. Santos · L. G. Fidalgo · R. P. Queirós · I. Delgadillo · J. A. Saraiva (✉)
QOPNA, Department of Chemistry, University of Aveiro, Aveiro, Portugal
e-mail: jorgesaraiva@ua.pt

R. S. Inácio · S. A. Moreira · R. V. Duarte
QOPNA, Department of Chemistry, University of Aveiro, Aveiro, Portugal

CBQF/Faculty of Biotechnology, Catholic University of Portugal, Porto, Portugal

A. M. P. Gomes
CBQF/Faculty of Biotechnology, Catholic University of Portugal, Porto, Portugal

S. Roohinejad et al. (eds.), *Effect of Emerging Processing Methods on the Food Quality*,
https://doi.org/10.1007/978-3-030-18191-8_4

Thus, in this chapter, HPP application on different food products will be addressed, i.e., fruits, vegetables, fish and meat products, milk and cheese, and its effects on the nutritional, textural and sensorial properties will be discussed. In each sub-section, tables are presented to summarize the most important studies; being highlighted in the text the most pertinent results therein.

4.2 Advantageous and Challenges of Using High-Pressure Processing on Food Quality

4.2.1 Nutritional Properties

4.2.1.1 Fruits and Vegetables

Fruits

Fruit is normally associated with great nutritional properties due to its high content of vitamins, carotenoids or polyphenols. HPP has appeared as a promising tool to process fruit products retaining most of its nutrients. For instance, both ascorbic acid (Tewari, Sehrawat, Nema, & Kaur, 2017) and phenolic compounds (Zhao, Zhang, & Zhang, 2017) have been shown to present higher stability after HPP and during storage of many fruit products, comparatively to those processed by traditional heat treatments. It is generally accepted that HPP has small effects on low molecular weight compounds such as vitamins C. In general, vitamin C is unaffected by HPP as most studies reported a retention above 80% after processing (Barrett & Lloyd, 2012), which can be maintained during 1–3 months at refrigerated storage (Tewari et al., 2017). For example, pressures between 350 and 600 MPa almost did not affect the vitamin C content of strawberry, blackberry (Patras, Brunton, Da Pieve, & Butler, 2009), tomato purées (Patras, Brunton, Da Pieve, Butler, & Downey, 2009), and orange juice (Sánchez-Moreno, Plaza, De Ancos, & Cano, 2003). However, there are still some exceptions, particularly at more severe conditions, both after processing and during storage. For instance, Landl, Abadias, Sárraga, Viñas, and Picouet (2010) reported higher losses of vitamin C at 600 MPa comparatively to 400 MPa after processing, and Valdramidis et al. (2009) reported losses during storage up to 36 days. There may be several reasons for different degradation rates, such as type of cultivar (Wolbang, Fitos, & Treeby, 2008), packaging and storage conditions, oxygen present and enzymatic activity (Valdramidis et al., 2009), however the most probable reason is due to oxidation (Oey, Van der Plancken, Van Loey, & Hendrickx, 2008). There are fewer studies regarding other vitamins, mainly vitamin B group, which are very stable under pressure and no major losses are reported after HPP (Barrett & Lloyd, 2012).

Similar to vitamins, carotenoids in fruits are not much affected by HPP and sometimes their availability even increases most likely due to the rupture of the cells of the fruit and leaking its content to the extracellular medium (Chen et al., 2013;

Sánchez-Moreno, de Ancos, Plaza, Elez-Martínez, & Cano, 2009). This effect was reported in several fruit products, for instance in tomato purée (600 MPa/15 min) (Patras, Brunton, Da Pieve, Butler, & Downey, 2009) and orange juice (400 MPa/1 min) (Plaza et al., 2011). In another study, Sánchez-Moreno et al. (2005) reported an increase in the availability of both total and individual carotenoids (β-cryptoxanthin, zeaxanthin, lutein, β-carotene and α-carotene). Nonetheless, there are some few contradictory results, such as those reported by Gupta, Kopec, Schwartz, and Balasubramaniam (2011), where a decrease of 25% in the availability of carotenoids in tomato juice after HPP was reported.

Similarly to vitamins and carotenoids, phenolic compounds on fruit products are minimally affected by HPP (Marszałek, Woźniak, Kruszewski, & Skąpska, 2017; Zhao et al., 2017). For instance, Chen et al. (2013) found that the availability of total phenols present in pomegranate juice increased after HPP, however the content of anthocyanins decreased from 68.54 mg/100 g before processing to 61.11 mg/100 g after processing. However, the impact was smaller than the traditional thermal processing, which decreased the content of anthocyanins to 59.51 mg/100 g of pomegranate juice. Similar results were obtained by Patras, Brunton, Da Pieve, and Butler (2009) in blackberry and strawberry purées. The authors reported an increase in the total phenols with increasing pressure (400–500 MPa) and an increase in anthocyanins at different HPP conditions except in the strawberry purée samples treated at 400 MPa. Overall, the general nutritional content of fruit-based products is well preserved by HPP and, in some cases, even improved.

Vegetables

HPP is mainly used to inactivate microorganisms and enzymes without degrading flavors/nutrients and minimizing the losses of beneficial ingredients in fruit and vegetable commodities (Pasha et al., 2014). Since HPP acts on volume compression, due to the low change in volume on low-molecular compounds, such as vitamins and functional content, the effects of this technology are expected to be minimum in these compounds (Wang, Huang, et al. 2016). The secondary metabolites produced by horticultural crops when they are subjected to environmental stresses are commonly known as bioactive compounds due to their health-promoting properties (Jacobo-Velázquez et al., 2017). When these stresses are applied during the postharvest phase, the biosynthesis and accumulation of these compounds can be enhanced by using HPP treatment (Dörnenburg & Knorr, 1998). HPP can also act as an extraction technique, enhancing the extractability of important compounds. For example, Paciulli, Medina-Meza, Chiavaro, and Barbosa-Cánovas (2016) have demonstrated that HPP applied at 650 MPa and room temperature (Table 4.1) can effectively increase the content of total phenolic compounds of beetroot, obtaining higher yields after longer extraction times (from 3 to 30 min), when compared to the raw vegetable. Similar results were obtained for red sweet pepper, another example of a nutritious food, being demonstrated that high pressures (above 500 MPa) can enhance the extractability of fiber (Hernández-Carrión et al., 2014). Analogous

Table 4.1 Examples of the main effects of HPP on the nutritional properties of vegetables

Vegetable products	HPP conditions	Nutritional effects	Reference
Beetroot slices	650 MPa/3–30 min/RT	Good retention of ascorbic acid after HPP when compared to thermal treatment. Higher extractability of total phenolic compounds after HPP treatment	Paciulli et al. (2016)
Spinach leaves	100, 250, 500, and 600 MPa/5 min/RT	HPP maintained the content of chlorophylls *a* and *b*	Wang, Ding et al. (2016)
Spinach, parsley, dill, kale	200, 400, 600 MPa/5, 10, 40 min/RT	HPP did not reduce the content in xanthophylls and chlorophylls; HPP allowed a higher extractability of these compounds in spinach purée	Arnold, Schwarzenbolz, and Böhm (2014)
Carrot, tomato, red pepper, broccoli, spinach, green pepper	625 MPa/5 min HPP: 20 °C HPHT: 70, 117 °C	While HPP treatment did not cause degradation on carotenoids and chlorophylls, HPHT process led to both chlorophylls' degradation	Sánchez, Baranda, and Martínez de Marañón (2014)
Onion, potato, pumpkin, red beet	600 MPa/15 min/117 °C	HPHT seems to reduce the formation of Maillard reaction and Strecker degradation products and enhance oxidative reaction products	Kebede et al. (2014)
Red sweet pepper	100–500 MPa/15 min/RT	Higher extractability of dietary fiber after HPP treatment at 500 MPa	Hernández-Carrión, Hernando, and Quiles (2014)
Carrot and spinach	100–500 MPa/20 min/20 °C	Good retention of ascorbic acid after HPP when compared to thermal treatment for both vegetables. Higher extractability of total phenols and flavonoids after HPP treatment	Jung, Lee, Kim, and Ahn (2013)

results were found after HPP treatment of carrot and spinach (Jung et al., 2013) since the total phenolic contents and flavonoids were the highest in HPP treated samples at 500 MPa, followed by thermal processing and HPP treated samples at 300 MPa, respectively. Total phenol and flavonoid contents were enhanced by increasing the pressure levels, which may be due to the enhanced membrane permeability (Jung et al., 2013).

Chlorophylls and chlorophyll-protein complexes are key factor quality in vegetable products. Wang, Ding, Hu, Liao, and Zhang (2016) showed that HPP under 100, 250 and 500 MPa (for 5 min at room temperature) allowed better retention of chlorophylls *a* and *b* of spinach leaves (Table 4.1), and the treatment at 100 MPa was reported to be most effective. Contrasting results were obtained for the thermal-treated samples at 100 °C for 60 s, where was found a significant ($p < 0.05$) decrease of the chlorophylls content, since the high temperature induced the disruption of the thylakoid membrane, while HPP was able to maintain a compact and stacked structure, similar to the untreated samples (Wang, Ding et al. 2016). Similar results were found by Arnold et al. (2014) who studied the effect of HPP at 200, 400, and 600 MPa, for 5, 10, and 40 min at room temperature on the bioactive compounds of spinach, parsley, dill, and kale (Table 4.1), and concluded that HPP maintained the concentration of xanthophylls (e.g. lutein and zeaxanthin) and carotenoids, compared to the samples subjected to heat for 5–20 min at 121 °C. In addition, HPP led to a significant higher extractability of these compounds regardless of the pressure level or the holding time, when compared to untreated spinach purée (Arnold et al., 2014).

Ascorbic acid is known for being easily degraded at high temperatures. After thermal treatment of carrot and spinach, it was found a decrease of 15% and 24% of ascorbic acid compared to the controls, respectively (Jung et al., 2013). Nevertheless, when the samples were treated by HPP (100–500 MPa, 20 min, 20 °C), good retention of ascorbic acid was reported when compared to the thermal treatment and control for both samples (Jung et al., 2013). Similar results were obtained by Paciulli et al. (2016), who reported better retention of ascorbic acid using HPP (650 MPa at room temperature) compared to thermal blanching.

4.2.1.2 Meat Products

Fish and Fish-Based Products

Fish has high nutritional value, especially high biological value proteins and lipids, being marketed and consumed worldwide. Additionally, fatty fish has a high concentration of n-3 fatty acids, which is often recognized by consumers as beneficial for human health (Ruxton, Calder, Reed, & Simpson, 2005). The effects of HPP on the fatty acid profile of HPP-treated fish were evaluated and no changes in the composition of this lipid fraction in turbot (Chevalier, Le Bail, & Ghoul, 2001) and coho salmon muscles (Ortea, Rodríguez, Tabilo-Munizaga, Pérez-Won, & Aubourg, 2010) were reported (Table 4.2). In another study, salmon processed by pressure resulted in the stable saturated, monounsaturated and polyunsaturated fatty acids, only with a decrease of n-6 fatty acids in samples treated at 300 MPa/15 min and consequent reduction of the n-6/n-3 ratio (Yagiz et al., 2009). On the other hand, Sequeira-Munoz et al. (2006) reported an increase of the free fatty acids content in carp muscle when compared to the untreated samples, using a pressure level between 140 and 200 MPa (for 15–30 min at 4 °C). In contrast, the formation of free fatty acids after processing was reduced at 150–450 MPa at 20 °C for 0–5 min (Aubourg et al., 2013). Furthermore, HPP seems to not induce any effect on lipase enzyme, being observed a similar activity to untreated samples in most of the HPP-treated samples (Teixeira et al., 2013).

Table 4.2 Some publications with the main effects of HPP on free fatty acids of fish

Fish products	HPP conditions	Nutritional effects	Reference
Atlantic salmon	150 and 300 MPa/ 15 min/room temperature	Stable total saturated, monounsaturated, and n-3 and n-6 polyunsaturated fatty acids profiles	Yagiz et al. (2009)
Carp	140–200 MPa/ 15–30 min/4 °C	Increase in free fatty acids levels	Sequeira-Munoz, Chevalier, LeBail, Ramaswamy, and Simpson (2006)
Coho salmon	135–200 MPa/ 30 s/15 °C	No significant changes	Ortea et al. (2010)
Mackerel	150–450 MPa/ 0–5 min/20 °C	Reduction of the formation of free fatty acids	Aubourg, Torres, Saraiva, Guerra-Rodríguez, and Vázquez (2013)
Turbot	100–200 MPa/ 15–30 min/4 °C	No significant changes	Chevalier et al. (2001)

Medina-Meza, Barnaba, and Barbosa-Cánovas (2014) observed a stronger catalytic oxidation power using pressure levels up to 300 MPa, although higher oxidation levels were observed in other studies at lower pressures between 150 and 300 MPa (Amanatidou et al., 2000; Lakshmanan, Patterson, & Piggott, 2005; Sequeira-Munoz et al., 2006; Teixeira et al., 2014). However, the effects of HPP on lipid oxidation of fish muscles also varied significantly depending on many factors, such as applied pressure level/pressure holding time, fish species and the type of muscle (dark or white). For example, a treatment lower than 400 MPa for 20 min at room temperature showed a slight effect on lipid oxidation in cod muscle compared to untreated muscle (Angsupanich & Ledward, 1998), whereas carp and turbot muscles were more susceptible to lipid oxidation after HPP-treatment of 100 MPa for 30 min at 4 °C (Sequeira-Munoz et al., 2006) and 100 MPa for 15 min at 4 °C (Chevalier et al., 2001), respectively.

There are a few research studies evaluating the cholesterol oxide formation in fish after HPP treatment. Although the cholesterol concentration did not change in mackerel and herring muscles after HPP treatment at 600 MPa for 10 min, higher cholesterol oxidation was observed in mackerel muscles (Figueirêdo, Bragagnolo, Skibsted, & Orlien, 2015). The cholesterol oxidation might be due to the breakage of cell structures and exposure of phospholipids membrane (Medina-Meza et al., 2014).

Meat and Meat-Based Products

HPP has been described as a powerful tool for the development of new/improved food products (Huang, Wu, Lu, Shyu, & Wang, 2017). However, some issues have been found on fresh meat products when processed by high pressure. One of the concerns, behind color losses, is the possible nutritional value decrease of the products processed by this technology.

It is known that a pressure level between 300 and 600 MPa led to the lipid oxidation of several meat products, being its lipid content and fatty acid composition of phospholipids and free fatty acids modified (Sazonova, Galoburda, & Gramatina, 2017). As the minimization of saturated fat intake with a concomitant increase of polyunsaturated fats is currently recommended, the oxidation of the unsaturated fatty acids may lead to a decrease in the nutritional value of the product (Ma & Ledward, 2004). Thus, some researchers have applied antioxidant active packaging and/or used different antioxidants to inhibit lipid oxidation (Sazonova et al., 2017).

Several studies have reported the possible fresh meat nutritional changes promoted by HPP and the oxidation process. In fact, in McArdle, Marcos, Kerry, and Mullen (2010) work, some differences were observed in specific fatty acids of bovine (*M. pectoralis profundus*) meat, not being found an overall HPP effect on polyunsaturated/saturated fatty acids (PUFA/SFA) or omega 6/omega 3 (n-6/n-3) ratio; however, the processing temperature influenced the sum of SFA, monounsaturated (MUFA) and PUFA fatty acids, being 40 °C the processing temperature that led to a higher SFA and PUFA and lower MUFA compared to HPP at 20 °C.

Nonetheless, in He et al. (2012) work, the authors found that a pressure level above 350 MPa in pork muscle caused marked lipolysis of the intramuscular phospholipids with a corresponding increase in free fatty acids. Over storage (at 4 °C) on HPP samples (350 and 500 MPa), the phospholipid breakdown increased with time with an increase of the free fatty acids and TBAR values. Thus, when the fatty acid composition of phospholipids was studied it was verified that SFA and MUFA increased with time, the result of palmitic and oleic acid contents while the PUFA content decreased, mainly due to losses of linoleic, linolenic and arachidonic acids (He et al., 2012). In what concerns HPP effect on fatty acid composition of intramuscular total lipids, some minor but not significant differences in fatty acid composition were observed between non-treated and treated samples, and for fatty acid composition of triglycerides, no significant differences in the percentage of every single fatty acid or SFA, MUFA and PUFA between non-treated and pressurized samples were observed, with the exception of linoleic acid (C18:2) (He et al., 2012). In another study performed by Wang et al. (2013), a loss of PUFA on HPP samples of yak body fat was detected, resulting in less favorable fatty acid profiles over storage. In this case, cold storage (20 days) of samples treated at 400 and 600 MPa led to an increase of TBAR values and a decrease in the sensory quality with a decrease in the PUFA/SFA and n-6/n-3 ratios.

In what concerns ultrastructure and *in vitro* protein digestion, Kaur et al. (2016) studied it on bovine *longissimus dorsi* muscle meat and verified that HPP meats (pepsin-digested, 60 min) presented fewer proteins or peptides of high molecular weight than on untreated meat, probably due to the breakdown proteins, mainly at 600 MPa pressure.

Nonetheless, some concerns regarding migration of compounds from plastic packages into the meat product have been addressed since HPP is applied to the final product already packaged. As an example, the study of Rivas-Cañedo, Fernández-García, and Nuñez (2009) allowed to observe both HPP and plastic packaging material effects on fresh meats, being concluded that as most of the compounds in the plastic were lipophilic, the migration of compounds into the meat could occur, being potentially harmful for the human health. In Table 4.3 are listed some works where the HPP effect on meat-based products was studied concerning its nutritional properties.

4.2.1.3 Dairy Products

Milk

High-pressure processing (HPP), as a cold alternative process to thermal pasteurization, has been applied in dairy food with mainly the advantage of shelf-life extension (2–3 times) relativity to non-pasteurized products (Dhineshkumar, Ramasamy, & Siddharth, 2016).

In dairy food, HPP has been applied in milk, fresh cheese, ripened cheese, whey cheese, yogurt, ice cream, and butter. However, the application on milk for subsequent cheese making and the application directly to the pressed curd and/or during cheese ripening have been the main areas in a study conducted by Martínez-Rodríguez et al. (2012). The effects of HPP are related to the pressure intensity, the holding time under pressure and the ripening stage when applied in cheese.

Table 4.3 Examples of HPP effects on the nutritional properties of different meat-based products

Meat products	HPP conditions	Nutritional effects	Reference
Cooked ham	100, 300 and 600 MPa/ 5 min/RT	45% salt reduction on cooked ham (salt content reduced to a level of 1.1% NaCl by the replacement of 0.2% of NaCl with KCl and combined HPP at 100 MPa after tumbling)	Tamm, Bolumar, Bajovic, and Toepfl (2016)
Serrano ham	600 MPa/6 min/21 °C	Only 8 volatile compounds presented to be HPP affected (higher levels of methanethiol and sulfur dioxide in HPP-treated samples and higher levels of ethyl acetate, ethyl butanoate, ethyl 2-methylbutanoate, ethyl 3-methylbutanoate, dimethyl disulfide and dimethyl trisulfide in untreated samples)	Martínez-Onandi et al. (2017)
Sliced skin vacuum packed dry-cured ham	600 MPa/6 min/15–32 °C	HPP did not produce changes in fatty acids content, protein content nor antioxidant enzyme activities HPP did not inhibit lipases and phospholipases action over storage (50 days with light) since treated and non-treated samples presented losses of fatty acids from phospholipid fraction while fatty acids increased from the free fatty acid fraction	Clariana et al. (2011)

The use of HPP in the food industry has been of great interest as a possible alternative to heat treatments for food preservation and processing. HPP eliminates or reduces the use of heating, and so avoiding thermal degradation of food components, and the retention of natural flavors, colors, and nutritional value (Huppertz et al., 2002). HPP has been applied to various food products, and several studies have been conducted for better knowledge on dairy foods, such as milk, cheese and yogurt (Ye, Anema, & Singh, 2004).

It has been reported that, among other effects, the application of high pressures to milk reduces its light-scattering properties, leading to a change in its appearance, and this was attributed to casein micelles disintegration into smaller structures (López-Fandiño, Fuente, Ramos, & Olano, 1998). In this study where bovine, caprine and ewe's milk were processed using 100–400 MPa for 5–30 min (Table 4.4),

Table 4.4 Examples of the main effects of HPP on the nutritional properties of milk

Dairy products	HPP conditions	Nutritional properties	Reference
Milk (bovine)	100, 200, 300 and 400 MPa/5–30 min/20 °C	Ca, P and Mg solubilization increased with pressurization up to 300 MPa, and then decreased slightly to 400 MPa. Maximum micelle dissociation was observed in milk treated at 300 MPa	López-Fandiño et al. (1998)
Milk (ovine)	100, 200, 300 and 400 MPa/5–30 min/20 °C	Soluble Ca, P and Mg increased with pressure but were smaller than those found in bovine milk. Milk dissociation increased with pressure up to 400 MPa	
Milk (caprine)	100, 200, 300 and 400 MPa/5–30 min/20 °C	Soluble Ca, P and Mg increment were more pronounced than those in the milk from the other two species. Micelle dissociation was observed at a maximum at 300 MPa	
Milk (bovine)	100–800 MPa/Up to 60 min/20 °C	A slight increase in fat globule size was observed with increasing pressure up to 700 MPa Association of β-LG with the milk fat globule occurred at pressures >100 MPa, while associations of α-LA and κ-casein occurred only at pressures ≥700 and 500 MPa, respectively	Ye et al. (2004)
Milk (ovine)	100–500 MPa/10 and 30 min/25 and 50 °C	Pressurization showed a tendency to increase milk fat globules in the range 1–2 μm	Gervilla, Ferragut, and Guamis (2001)
Milk (human)	400, 500, and 600 MPa/5 min/12 °C	HPP allowed better maintenance of the vitamin C and the same levels of preservation of fatty acid proportions and tocopherols when compared to thermal pasteurization	Moltó-Puigmartí, Permanyer, Castellote, and López-Sabater (2011)

the release of micellar Ca, P, and Mg into the serum was observed (López-Fandiño et al., 1998). Changes in the content of soluble Ca, P, and Mg were more pronounced in ewe's milk processed by HP when compared to bovine and caprine milk. Unlike bovine and caprine milk, pressurization at 400 MPa increased the levels of these three elements in the serum over the treatment at 300 MPa. In the case of bovine and caprine milk, solubilization increased with pressurization up to 300 MPa and then decreased slightly to 400 MPa. This can be related to micelle dissociation, since in the case of bovine and caprine milk, maximum dissociation from the micelle was found in milk treated at 300 MPa, while in ewes' milk, dissociation increased with pressure up to 400 MPa (López-Fandiño et al., 1998). Ye et al. (2004) observed that, in general, micelle size is not affected by treatments at 100 or 200 MPa for 20 min at 20 °C, while treatment at 250 MPa increases micellar size by 20%, and higher pressures (300–800 MPa) decrease it by 50%. The associations of α-LA and κ-casein occurred at pressures ≥700 MPa and 500 MPa, respectively, but the amounts are lower than that observed for β-LG. *α*-LA has a more rigid molecular structure, being much more resistant than β-LG to denaturation during high-pressure treatment.

Moltó-Puigmartí et al. (2011) suggested the possibility to process human milk using HPP, since fatty acid proportions in milk, δ-, γ-, and α-tocopherols, total vitamin C and ascorbic acid levels where preserved, while on the other hand, thermal pasteurization lead to a reduction in the nutritional content of human milk.

Fresh Cheese

Fresh cheese is a dairy product that has a short shelf life due to its pH near to neutral (no starter is added), high moisture and high-water activity (>0.99), which allows a fast growth of spoilage microorganisms. HPP could be a solution for minimal processing this highly nutritious and perishable food product while restraining microbial growth.

Overall, HPP resulted in fresh cheeses whey loss, which contributed to a reduction in moisture content (Table 4.5). For example, Evert-Arriagada et al. (2012) reported a reduction on microbial growth and no changes in fat and total protein contents or pH values, but significant whey loss occurred at 400 MPa for 5 min. Similar results were observed by the same authors using a higher pressure (500 MPa for 5 min), with a significant increase ($p < 0.05$) in whey loss from day 14 and the highest measure of free whey at day 21, with the highest amount of total solids observed at the same sampling day (Evert-Arriagada et al., 2014). On the other hand, Okpala et al. (2010) reported that moisture content did not vary from treatments using pressure up to 150 MPa, but dropped more when the pressure increased over 150 MPa.

Table 4.5 Examples of the main effects of HPP on the nutritional properties of fresh cheese

Dairy products	HPP conditions	Nutritional properties	References
Fresh Cheese (cow)	200, 400, and 600 MPa/3–20 min/20 and 40 °C	A small decrease in the protein content HPP resulted in moisture content reduction, as pressure increased (up to 2%)	Van Hekken, Tunick, Farkye, and Tomasula (2013)
Fresh Cheese (cow)	500 MPa/5 min/16 °C	The total solids content increased ($p < 0.05$) within the last 7 days of storage for pressurized samples No statistical variations were observed on whey loss after pressurization	Evert-Arriagada, Hernández-Herrero, Guamis, and Trujillo (2014) and Van Hekken et al. (2013)
Fresh Cheese (cow)	400 MPa/20 min/20 °C	HP cheese presented pH values higher than the control cheese	Sandra, Stanford, and Goddik (2004)
Fresh Cheese (cow)	300 or 400 MPa/5 min/6 °C	The total solid content of HP cheeses was higher than control cheeses, with no changes on fat and total protein contents, or pH values Significant whey loss was observed on cheeses treated at 400 MPa	Evert-Arriagada, Hernández-Herrero, Juan, Guamis, and Trujillo (2012)
Fresh Cheese (cow)	9, 50, 150, 250, and 291 MPa/1, 5, 15, 25, and 29 min/25 °C	Treatments over 150 MPa resulted in a more significant reduction in moisture content, and an increment of fat content The protein content of HPP fresh cheese in all treatments remained lower than the control pH and TBA value (lipid oxidation) decreased with increasing pressure	Okpala, Piggott, and Schaschke (2010)
Fresh Cheese (goat)	500 MPa/5, 15 and 30 min/10 or 25 °C	Non-protein nitrogen of pressurized cheeses was lower than that of control cheeses Pressure-treated cheeses expelled significantly more whey than control cheeses	Capellas, Mor-Mur, Sendra, and Guamis (2001)

Cheese

HPP has been applied to cheese in order to increase the preservation and/or the modification of the ripening process (deceleration/acceleration) (Martínez-Rodríguez et al., 2012; O'Reilly, Kelly, Murphy, & Beresford, 2001; Trujillo, Capellas, Saldo, Gervilla, & Guamis, 2002). However, in any HPP application, it is important to analyze the effect on nutritional properties. As shown in Table 4.6, HPP may cause or not lower effect on moisture, protein and fat content when applied in cheese with some days of ripening (Calzada, del Olmo, Picon, Gaya, & Nuñez, 2014a; Delgado et al., 2012, 2015; Voigt et al., 2010). When the treatment was applied in cheese curd (Calzada, del Olmo, Picon, Gaya, et al., 2014b, c; Delgado et al., 2012; Saldo et al., 2002; Voigt et al., 2010), major differences in nutritional composition occurred after HPP and during its ripening. In cheese, a relevant nutritional property is a proteolysis, which indicates the age of the cheese. In general, some studies revealed proteolysis acceleration when cheeses were treated by HPP with few days of ripening (Garde et al., 2007; Saldo et al., 2003; Voigt et al., 2010). Other researchers observed a deceleration or same proteolytic indexes of ripened cheeses treated by HPP (Calzada, del Olmo, Picon, Gaya, & Nuñez, 2014a; Garde et al., 2007). The lipolysis of short-chain (SC-), medium chain (MC-) and long-chain free fatty acids (LC-FFA) showed no significant changes after HPP treatment (Calzada et al., 2015; Calzada, del Olmo, Picon, Gaya, et al., 2014c; Voigt et al., 2010).

4.2.2 *Textural Properties*

4.2.2.1 Fruits and Vegetables

Fruits

The texture of fruit products is of paramount importance for the consumer acceptability. In a general way, HPP may decrease the hardness and chewiness in fruits. An example is the work of Denoya, Vaudagna, and Polenta (2015) who studied vacuum packed fresh-cut peaches subjected to HPP (500 MPa/5 min/20 °C) and their texture profile, that was analyzed for hardness, cohesiveness, and chewiness. The authors reported a decrease in hardness and chewiness after HPP. Furthermore, according to Miguel-Pintado et al. (2013), the intensity of HPP is also significant in what concerns texture modification, since nectarine halves subjected to 200–300 MPa presented fewer firmness changes than those subjected to 600 MPa. Likewise, hardness was reduced in a number of fruits following HPP, such as persimmons (Vázquez-Gutiérrez, Quiles, Hernando, & Pérez-Munuera, 2011), pumpkin (Zhou et al., 2014), and strawberry (Gao et al., 2016). The observed changes in the textural properties reported in these studies are most likely the result of the effects of HPP on the microstructure of fruits, namely the destruction of cell walls

Table 4.6 Effect of HPP on nutritional properties of different cheeses

Dairy products	HPP conditions	Textural effects	Reference
Casar Cheese (raw ewe milk)	200 or 600 MPa/5 or 20 min/10 °C	Moisture content: HPP causing no changes (at 60 days of ripening)	Delgado, Rodríguez-Pinilla, Márquez, Roa, and Ramírez (2015)
Ibores Cheese (raw goat milk)	400 or 600 MPa/7 min/10 °C	Fat content: HPP causing no changes (at 2 and 31 days of ripening) Protein content: HPP causing no changes (at 2 and 31 days of ripening), except to HPP at 600 MPa/7 min treated cheeses (at 50 days of ripening) with lower content	Delgado, González-Crespo, Cava, and Ramírez (2012)
Garrotxa cheese (pasteurized goat milk)	400 MPa/5 min/14 °C	Moisture content: higher for HHP cheeses (at 4 days of ripening)	Saldo, McSweeney, Sendra, Kelly, and Guamis (2002)
		Proteolytic indexes: higher for HPP cheeses (at 4 days of ripening)	Saldo et al. (2003)
Brie cheese (pasteurized cow milk)	400 or 600 MPa/5 min/9–14 °C	Moisture content: did not vary immediately after HPP (at 21 days of ripening)	Calzada, del Olmo, Picon, Gaya, et al. (2014b)
		SC.FFA: higher on HPP MC- and LC-FFA: HPP cheeses showed equal content when treated at 14 days and higher content were treated at 21 days	Calzada, del Olmo, Picon, and Nuñez (2014c)
NA (raw cow milk)	400 or 600 MPa/5 min/14 °C	SC-, MC-, LC-FFA: HPP cheeses showed similar concentration to control cheeses (21 days or ripening)	Calzada, del Olmo, Picon, and Nuñez (2015)
La Serena (raw ewe milk)	300 or 400 MPa/10 min/10 °C	Proteolysis: higher when HHP treatments were applied at 400 MPa on d 2 compared to other treatments	Garde, Arqués, Gaya, Medina, and Nuñez (2007)
Irish blue-veined cheese	400 or 600 MPa/10 min/20 °C	Moisture, fat and protein content: small differences between HPP and control cheeses Proteolytic indexes: higher for HHP cheeses treated Lipolysis/FAA: no significant effect of HPP (at 42 days of ripening)	Voigt, Chevalier, Qian, and Kelly (2010)

and the dispersion of intracellular component throughout the tissue (Vázquez-Gutiérrez et al., 2011), as well as due to the activity of the pectin methylesterase, facilitated by the release of the cell-wall-bonded enzyme and the contact with its substrate, which causes the de-esterification of pectin thus changing the fruit texture (Basak & Ramaswamy, 1998). Therefore, it is expected a decrease in hardness during storage due to enzymatic and non-enzymatic depolymerization of pectins, as seen, for example, in the work of Gao et al. (2016), and Zhou et al. (2014), who reported a decrease in hardness after storage at 4 °C for 45 and 60 days, respectively. Still, HPP retains textural properties at acceptable levels.

Zhang et al. (2012) reported that the hardness of yellow peach was preserved by 69.5% after 3 months of refrigerated storage after HPP (600 MPa/5 min). Furthermore, HPP seems to better retain textural properties than thermal processing. Gao et al. (2016) described a higher hardness of the flesh of strawberries after HHP (400 MPa/5 min) than after thermal processing (75 °C/20 min). Similar results were described in yellow peaches, most likely due to the higher structural damages caused by thermal processing comparatively to HPP Zhang et al. (2012).

Vegetables

Food texture can change after processing and affect the quality and acceptance of the product by the consumer. Thermal treatment is known to affect vegetable texture and structure, mainly due to loss of instrumental firmness by membrane disruption.

After HPP treatment (650 MPa at room temperature, for 3–30 min) of beetroot, it was demonstrated that high pressure allowed maintaining the vegetable's cut hardness similar to the raw untreated samples after 30 min of processing (Paciulli et al. 2016). It was also noteworthy that after long processing times, a higher recovery of consistency was found compared to short-time processed samples, being these results explained by the possible tissue recovery during the holding time by fortification of intercellular adhesion, since formation of new ionic linkages occurs in cell wall pectic polysaccharides (Paciulli et al., 2016; Trejo Araya et al., 2007). Other parameters such as hardness, chewiness, and cohesiveness were studied, being the results consistent with the previous, since the samples pressurized for 15 and 30 min presented similar characteristics to the raw sample, while the beetroot processed for 3 and 7 min presented lower values (Paciulli et al., 2016). Similar results were presented by Hernández-Carrión et al. (2014) who showed that the higher the pressure used (100–500 MPa, for 15 min, at room temperature), the least impact on the microstructure was found on red sweet pepper tissue (Table 4.7).

Turgor is an important textural characteristic since it has a direct impact on the fresh appearance of the vegetables. The effect of HPP on cell turgor of red cabbage was studied by Rux et al. (2017) at pressures between 150 and 250 MPa, for 5–20 min, at 35–55 °C (Table 4.7). The main results focused on the fact that cell turgor can be recovered within hours if the pressure treatment was not too strong (150 MPa up to 10 min). Nonetheless, when the pressure level was above 175 MPa, and the temperature above 45 °C, the cells were irreversibly damaged (Rux et al., 2017).

Table 4.7 Examples of the main effects of HPP on the textural properties of vegetables

Vegetable products	HPP conditions	Textural effects	Reference
Red cabbage leaves	150–250 MPa/5–20 min/35–55 °C	Cell turgor affected by HPP above 150 MPa and temperature above 45 °C. Above these thresholds occur irreversible turgor losses	Rux, Schlüter, Geyer, and Herppich (2017)
Beetroot slices	650 MPa/3–30 min/RT	Higher cut hardness, recovery of consistency, chewiness, and cohesiveness for the samples treated by HPP for longer times (15 and 30 min)	Paciulli et al. (2016)
Asparagus spears	10–600 MPa/0.5–30 min/RT	HPP resulted in a decreased firmness	Yi et al. (2016)
Cocoyam, carrot, sweet potato	600 MPa/5, 30 min/RT	Reduced maximum cutting force and lower rigidity for HPP samples	de Oliveira, Tribst, Leite Júnior, de Oliveira, and Cristianini (de Oliveira, Tribst, Leite, de Oliveira, & Cristianini, 2015)
Red sweet pepper	100–500 MPa/15 min/RT	Low effect on cell's microstructure after HPP treatment at 500 MPa	Hernández-Carrión et al. (2014)

Contrasting results were reported by de Oliveira et al. (2015) after HPP processing of cocoyam, Peruvian carrot, and sweet potato. These authors reported that HPP has reduced the maximum cutting force and lower rigidity (up to 25%) compared to the control sample, when high pressures (600 MPa) and longer holding periods (up to 30 min) were used, being these results attributed to the partial starch gelatinization that occurs in these vegetables (de Oliveira et al., 2015).

4.2.2.2 Meat Products

Fish and Fish-Based Products

In fish texture profile analysis, the hardness, and springiness were mainly investigated, but cohesiveness, gumminess, adhesiveness, and chewiness were also studied. Table 4.8 presents some publications with the main results of the HPP application in fish texture.

An increase of muscle hardness was observed in HPP-treated cod (Montiel et al., 2012), trout and mahi-mahi (Yagiz et al., 2007), salmon (Yagiz et al., 2009), and tuna (Ramirez-Suarez & Morrissey, 2006). Despite HPP-treated sea bass (100 and 300 MPa for 5 min at 10 °C) exhibits lower hardness than the control, similar values were observed when the samples were subjected to 400 and 500 MPa for 5 min at 10 °C (Chéret et al., 2005).

Table 4.8 Some publications with the main effects of HPP on the texture properties of fish

Fish products	HPP conditions	Textural effects	Reference
Cod	400–600 MPa/5 and 10 min	The increase in hardness and shear strength	Montiel, De Alba, Bravo, Gaya, and Medina (2012)
Mahi mahi	150–600 MPa/ 15 min/RT	The lowest hardness was observed on control samples The increase of chewiness and gumminess with the increase of pressure	Yagiz, Kristinsson, Balaban, and Marshall (2007)
Rainbow trout	150–600 MPa/ 15 min/RT	The increase of hardness at 450 and 600 MPa, compared to 0.1 and 150 MPa The increase of cohesiveness after HPP	Yagiz et al. (2007)
Salmon	150 and 300 MPa/ 15 min	The increase of hardness, gumminess, and chewiness, and a decrease in adhesiveness	Yagiz et al. (2009)
	100–200 MPa/ 10–60 min/ 1 and 5 °C	The increase of cutting strength at 150 and 200 MPa for 30 min or 200 MPa for 60 min	Amanatidou et al. (2000)
Sea bass	100–500 MPa/5 min/10 °C	The decrease of hardness at 100–300 MPa, and no changes at 400 and 500 MPa No changes of cohesiveness, springiness, and resilience The decrease of gumminess and chewiness at 100–300 MPa and the increase at 400 and 500 MPa	Chéret, Chapleau, Delbarre-Ladrat, Verrez-Bagnis, and de Lamballerie (2005)
Tuna	275 and 310 MPa/ 2, 4 and 6 min	The increase of hardness with pressure and holding time increase	Ramirez-Suarez and Morrissey (2006)

Several possible explanations were suggested for the increase in fish muscles hardness under HPP. According to Angsupanich and Ledward (1998), the unfolding of actin and sarcoplasmic proteins and the formation of new hydrogen-bonded networks could contribute to the increase in hardness and springiness of pressurized fish muscles.

These changes in fish texture under HPP are linked with protein modifications, mainly the interactions between actin and myosin, the release of α-actinin, and the denaturation of myofibrillar proteins. Collagen is very stable at high pressure (Guyon, Meynier, & de Lamballerie, 2016).

Meat and Meat-Based Products

Although HPP could be an interesting food processing technology for food preservation, in what concerns to meat products it is capable to modify its properties, for instance, changing meat texture as a free-additive and physical process to tenderize and soften meat and meat products. In fact, although dependent on pressure level, temperature, pH, and ionic strength, HPP effect on sarcoplasmic proteins (mainly enzymes and heme pigments) can lead to the denaturation (mainly pressures above 200 MPa) promoting changes on water holding capacity, color, and myofibrillar proteins, being the latter's strongly related to the meat structure, and unfolded above 300 MPa (Marcos, Kerry, & Mullen, 2010; Sazonova et al., 2017; Sun & Holley, 2010). On the other hand, as the triple helix of collagen is predominantly stabilized by hydrogen bonds, it is expected to be inert to pressure in normal conditions, being this crucial for meat texture since connective tissue (mainly collagen) and contractile systems (mainly actomyosin) play an important role (Ma & Ledward, 2013).

Several studies have been carried out concerning the effect of HPP on meat texture, for instance, Morton et al. (2017) concluded that when two levels of pressure were applied (to *longissimus thoracis* (strip loin) steaks) within 1 h of slaughter and chilled for 1 day before freezing, HPP allowed to obtain meat after 1 day of storage with 60% lower shear force and higher sensory eating quality scores when compared to the non-processed product. It was also observed that HPP increased the final pH and decreased cooking loss (Morton et al., 2017).

In fact, since 1973, HPP (<138 MPa) is reported as a technology capable to produce substantial improvements in pre-rigor meat tenderness (Macfarlane, 1973). Since then, several works allowed to conclude, for instance, that pressure could cause a shortening of ≥30% but on cooking little further shortening occurred compared to unprocessed samples, weep and cooking losses could be similar but when HPP samples were cooked presented slightly higher moisture contents, cooked HPP samples were significantly more tender than unprocessed samples as judged by taste panels and Warner Bratzler Shear values, and taste panels rated pressure-treated samples more acceptable than the untreated ones (Ma & Ledward, 2013).

Although HPP effect on pre-rigor is highly dependent on several conditions, as muscle temperature, pressure level, processing time, among others, it can be used to improve water holding capacity of meat (and meat-based products) replacing the need to use additives (salt/phosphates) or non-meat ingredients (polysaccharides/non meat proteins) (Ma & Ledward, 2013). For instance, in Souza et al. (2011) study regarding HPP of pork pre-rigor carcasses, it was concluded that this technology partially inhibited the post-mortem metabolism leading to better water-holding capacity parameters (drip loss and cook loss), being observed that HPP improved pork palatability parameters, where Warner-Bratlzer shear force values indicated an increase of mechanical tenderness (also confirmed by sensory evaluation of tenderness), and also no changes promoted by HPP on collagen were detected (Souza et al., 2011).

In what concerns to post rigor fresh meat, it was already observed that a pressure treatment (20 min) at room temperature promoted myofibrillar proteins denaturation, leading to the toughness of the meat, mainly up to 400 MPa. However, when HPP was applied at 60–70 °C up to 200 MPa, significant decreases in hardness, gumminess, and chewiness were detected (Ma & Ledward, 2004). In this study, as expected, collagen showed an inert behavior to pressure (unfolded only at temperatures of 60–70 °C), and myosin revealed to be easily unfolded by both pressure and temperature, being formed new and modified structures of low thermal stability when pressure denature the initial structures (Ma & Ledward, 2004).

It has to be noted that enzymes are also important on meat texture properties, and for so, HPP effect on enzymes activity must be carefully analyzed since it can be catalyzed or inhibited by pressure in the majority of cases with its denaturation (Ma & Ledward, 2013). In their work, Ma and Ledward (2004) suggested that probably the application of 200 MPa at high temperature accelerated proteolysis, being this fact the major cause for the hardness decrease. Other studies performed at moderate pressure and high temperatures revealed similar results on meat hardness (Beilken, Macfarlane, & Jones, 1990; Zamri, Ledward, & Frazier, 2006). One possible explanation for this behavior could be explained by the temperature increase up to the set temperature (including the adiabatic heating induced by HPP) where enzymes still active and the pressure combined with slowly rising temperature could increase their reaction rate, inducing protein structure modification, so that marked proteolysis can occur (Ma & Ledward, 2013).

Concerning specific enzymes, it was also shown that HPP can increase cathepsin activity and decrease calpastatin, an important inhibitor of calpains, being also the calpains pressure sensitive and their activity decreased, for instance at 250 MPa (over 10 min at room temperature) (Sikes, Tornberg, & Tume, 2010; Sikes & Warner, 2016). Regarding cathepsins (pressure resistant up to 500 MPa), as they are located in the lysosomes, and these rupture around 200 MPa, they are released into the cytoplasm at these pressures (Homma, Ikeuchi, & Suzuki, 1994). Thus, Sikes et al. (2010) suggested that the catheptic degradation at the Z-line (where the actin has been depolymerized) is a necessary precursor for the establishment of a strengthened but brittle myofibrillar network on meat. Some works regarding the effect of HPP on textural properties of meat-based products are presented in Table 4.9.

Table 4.9 Examples of HPP effects on the textural properties of different meat-based products

Meat products	HPP conditions	Textural effects	Reference
Meat batters and reduced-fat sausages	200 MPa/ 2 min/10 °C	Textural and rheological properties improvement. HPP sausages with 20% fat presented similar cooking losses and texture results when compared to AP sausages with 30% of fat	Yang et al. (2016)
Pork sausages containing carrot dietary fiber	500 and 600 MPa/ 1 s, 3, 6, and 9 min/ 40, 50, and 60 °C	Young's Modulus increased with HPP treatments and affected Hencky strain values. HPP and carrot dietary fiber improved emulsion strength resulting in firm sausages	Grossi, Søltoft-Jensen, Knudsen, Christensen, and Orlien (2011)
Low-acid fermented sausages	400 MPa/ 10 min/17 °C	No differences were detected between non-treated and HPP treated sausages with the exception of an increase in textural properties (higher cohesiveness, chewiness, and springiness)	Marcos, Aymerich, Dolors Guardia, and Garriga (2007)

4.2.2.3 Dairy Products

Milk

HPP of milk resulted in some cases in the increase of milk viscosity as pressure increases (Harte, Luedecke, Swanson, & Barbosa-Cánovas, 2003; Huppertz et al., 2002; Trujillo et al., 2007) as detailed in Table 4.10. Increase in the viscosity could be related to the disintegration of the casein micelles into smaller structures and the denaturation of β-LG, which could produce large protein aggregates leading to increasing the milk viscosity (Trujillo et al., 2007). Also, changes in viscosity could be related to the liberation of colloidal calcium phosphate and individual caseins concentrating in the serum in which submicelles are suspended (Harte et al., 2003).

Fresh Cheese

Fresh cheeses processed by HPP resulted in several textural changes, which were pressure intensity dependent (Table 4.11). Van Hekken et al. (2013) reported that heating fresh cheese prior to HPP affected fresh cheese texture (40 °C/400 MPa for 20 min or at 600 MPa for 5, 10, or 20 min) resulting in the highest fracture rigidity ($p < 0.05$), while fresh cheeses processed at 20 °C had less variation among treatments and were closest to the non-processed ones. Evert-Arriagada et al. (2012) observed that HPP fresh cheeses were significantly firmer than the control ones, and it could be related to lower water content compared to control cheeses. Similar results were described in other studies, where pressurized cheeses were more resistant to deformation (higher modulus values), and less fracturable and deformable than the control cheeses (Capellas et al., 2001; Evert-Arriagada et al., 2012; Sandra et al., 2004).

Table 4.10 Examples of the main effects of HPP on the textural properties of fresh milk

Dairy products	HPP conditions	Textural effects	Reference
Milk (bovine)	100 to 600 MPa/Up to 60 min/20 °C	The viscosity of skimmed milk increased with, up to a value of 2.5 as pressure and treatment time increased	Huppertz et al. (2002)
Milk (bovine)	310 MPa/0.3 s	Continuous high-pressure throttling increased the viscosity	Adapa, Schmidt, and Toledo (1997)
Colostrum (caprine)	400 and 500 MPa/10 min/20 °C	Samples processed with 500 MPa, presented higher visual viscosity	Trujillo et al. (2007)
Milk (bovine)	300, 400, 500, and 676 MPa/5 min/4 °C	An increase in milk viscosity after HPP was observed	Harte et al. (2003)

Table 4.11 Examples of the main effects of HPP on the textural properties of fresh cheese

Dairy products	HPP conditions	Textural effects	Reference
Fresh Cheese (cow)	200, 400, 600 MPa/3–20 min/20 and 40 °C	HPP cheeses tended to present higher values for hardness, chewiness, cohesiveness, fracture stress, and fracture rigidity	Van Hekken et al. (2013)
Fresh Cheese (cow)	500 MPa/5 min/16 °C	Pressurized cheeses were more resistant to deformation, less fracturable and deformable, than control cheeses	Evert-Arriagada et al. (2014)
Fresh Cheese (cow)	400 MPa/20 min/20 °C	HP cheese had higher firmness, gumminess, and chewiness than the control ones	Sandra et al. (2004)
Fresh Cheese (cow)	300 or 400 MPa/5 min/6 °C	In general, HP cheeses were significantly firmer than control cheeses	Evert-Arriagada et al. (2012)
Fresh Cheese (cow)	9, 50, 150, 250, and 291 MPa/1, 5, 15, 25 and 29 min/25 °C	Increased pressures led to increased hardness, but decreased adhesiveness of the HP-treated fresh cheese	Okpala et al. (2010)
Fresh Cheese (goat)	500 MPa/5, 15 and 30 min/10 or 25 °C	Fracture stress values were significantly higher when compared to control cheeses	Capellas et al. (2001)

Cheese

In general, HPP ripened cheeses revealed lower values of hardness, adhesiveness, firmness, elasticity, gumminess, and chewiness compared to control cheeses (Calzada, del Olmo, Picon, Gaya, et al., 2014b; Calzada, del Olmo, Picon, & Nuñez, 2014c; Delgado et al., 2012). However, after some days of storage, HPP cheeses recovered the textural properties to values closer to control cheeses (Calzada, del Olmo, Picon, Gaya, et al., 2014b) (Table 4.12).

4.2.3 *Sensorial Properties*

4.2.3.1 Fruits and Vegetables

Fruits

The nutritional compounds mentioned above play an important role in the general sensorial properties of food products, and since most of these compounds are minimally affected by HPP, also the sensorial properties are minimally affected.

One of the most studied sensorial parameters is the color; usually using the CIE 1976 ($L*$, $a*$, $b*$) color space parameters. Overall the color of fruit products is not

Table 4.12 Effect of HPP on textural properties of different cheeses

Dairy products	HPP conditions	Textural effects	Reference
Casar Cheese (raw ewe milk)	400 or 600 MPa/ 5 min/14 °C	Firmness and elasticity: higher values for the 600 MPa cheese and lower values for the 400 MPa cheeses in comparison to control cheeses (at 35 days of ripening)	Calzada, del Olmo, Picon, and Nuñez (2014a)
Brie cheese (pasteurized cow milk)	400 or 600 MPa/ 5 min/9–14 °C	Fracturability, elasticity, and firmness: decreased immediately after HPP (at 21 days of ripening), then HPP cheese revealed similar or higher texture characteristics than control	Calzada, del Olmo, Picon, Gaya, et al. (2014b)
Ibores Cheese (raw goat milk)	400 or 600 MPa/ 7 min/10 °C	Hardness, adhesiveness, gumminess, and chewiness: decreased after HPP (at 1, 30 and 50 days of ripening) Cohesiveness and springiness: was kept after HPP (at 1 and 30 days of ripening)	Delgado et al. (2012)
Commercial cheese (pasteurized ewe milk)	300 MPa/10 min	Fracture stress: increased immediately after HPP at 1 day of ripening and decreased when treated after 25 d of ripening Fracture strain: increased immediately after HPP	Juan, Ferragut, Guamis, and Trujillo (2008)

much affected by HPP rendering ΔE values generally below 4 (Koutchma, Popović, Ros-Polski, & Popielarz, 2016). Some recent examples are: i) the work of Chang, Wu, Chen, Huang, and Wang (2017) with grape juice where HPP, using 300 and 600 MPa for 3 min at 20 °C, resulted in a ΔE below 1 after treatment and up to 20 days of storage; ii) the work of Yi et al. (2017) that studied apple juice from 3 varieties and where HPP (600 MPa/3 min/10 °C) resulted in ΔE values below 2.5 (Pink Lady = 2; Granny Smith = 0.8; Jonagold = 2.5). Still, there are some cases in which the color changes were more noticeable. Among them, some studies stand-out, for example, studies with blood orange, watermelon, and orange juices, with ΔE values of 4.5, 8.0 and 9.3, respectively. It is noteworthy to mention that these juices were processed at severe conditions (600 MPa), particularly the watermelon juice that was processed at more extreme conditions than most fruit products usually are, namely 900 MPa for 60 min at 60 °C (Hartyáni et al., 2011; Torres et al., 2011; Zhang et al., 2011).

HPP has proven to also maintain other sensorial properties, considering several descriptors for acceptability such as odor and taste. Picouet et al. (2016) studied the effects of 350 MPa for 5 min on a multi-fruit smoothie containing apples, strawberries, oranges, and bananas. The authors compared the effect of HPP and thermal processing on odor, general appearance, and mouthfeel, and concluded that the HPP samples were similar in the overall sensory analysis to the untreated smoothies. However, the HPP-treated juices presented lower stability during storage, most likely due to the inefficacy of the process to considerably reduce the activity of oxidative enzymes. In another study with red fruit-based smoothies, similar results were obtained. Several parameters were evaluated, namely appearance, odor, flavor, and mouthfeel. According to the authors, moderate HPP conditions (350 MPa/10 °C/5 min) resulted in "fresh-like" smoothies, both after processing and during storage at 4 °C up to 14 days, presenting a non-altered sensory quality (Hurtado et al., 2017). Baxter, Easton, Schneebeli, and Whitfield (2005) studied several sensory properties of navel orange juice processed by HPP (600 MPa/1 min/20 °C) and its general consumer acceptability. The authors reported that the HPP juice had an odor and flavor that was acceptable by consumers, both after processing and up to 12 weeks of storage. Furthermore, the 20 key aroma components were analyzed by gas chromatography-mass spectrometry and it was concluded that the results of the HPP juice were similar to the untreated juice. Most studies available in the literature point to HPP as a good alternative for the stabilization of fruit-based products, since it is an effective technology maintaining good sensory qualities.

Vegetables

As HPP has a limited effect on the covalent bonds of low molecular weight compounds, such as flavor and color compounds of food products, it is expected that this technology will help to minimize the color changes and formation of off-flavors caused by thermal pasteurization (Wang, Huang et al. 2016).

A study on the color changes on cocoyam and carrot showed that HPP at 600 MPa, for only 5 min at room temperature, allowed to maintain the color unchanged, mainly due to enzyme inactivation, such as polyphenoloxidase (PPO) and peroxidase (Tribst, Leite, de Oliveira, & Cristianini, 2016). Also in cucumber juice, HPP treatment (at 500 MPa, for 5 min, at room temperature) seems to not have an immediate effect on juice color, since the CIE Lab parameters remained unchanged (Liu, Zhang, Zhao, Wang, & Liao, 2016) (Table 4.13). Nevertheless, after 20 days of storage, some clear changes were perceived, with high values of ΔE, but still, about 3.4-fold lower than the samples treated by high temperature (Liu et al., 2016). Similar results were described by Contador et al. (2014) who found that pumpkin purée treated at 400 MPa showed the same coloration parameters as non-processed purée, presenting an increase of about nine-folds of the ΔE parameter also after storage at refrigeration for 20 days (Contador et al., 2014). Contrasting results were reported by García-Parra et al. (2016), who related the instrumental color and PPO enzyme activity, after HPP treatment of pumpkin purée (Table 4.13). The authors have found that the purée lightness (CIE $L*$) was significantly higher after high pressure/mild temperature (600 MPa/60 °C) treatment. Also, the ΔE parameter was around 3–4, indicating a color difference perceptible by most consumers, even though that those conditions were the most effective to reduce PPO activity (García-Parra et al., 2016).

Table 4.13 Examples of the main effects of HPP on the sensorial properties of vegetables

Vegetable products	HPP conditions	Sensorial effects	Reference
Cocoyam and carrot	600 MPa/5 min/RT	Color parameters unchanged after HPP	Tribst et al. (2016)
Cucumber juice	500 MPa/5 min/RT	Color parameters unchanged after HPP	Liu et al. (2016)
Pumpkin purée	400, 600 MPa/ 5 min/10 °C	Treated samples (400 MPa) with same coloration parameters as non-processed purée	Contador, González-Cebrino, García-Parra, Lozano, and Ramírez (2014)
Pumpkin purée	300–900 MPa/ 1 min/60–80 °C	Higher lightness for samples treated by HPP at 600 MPa and 60 °C	García-Parra, González-Cebrino, Delgado, Cava, and Ramírez (2016)
Carrot purée	600 MPa/up to 180 min/118 °C	The formation rate of off-flavors was ten times slower than heated samples at 0.1 MPa	Kebede et al. (2017)
Onion, potato, pumpkin, red beet	600 MPa/ 15 min/117 °C	HPHT seems to reduce the degradation of Strecker products	Kebede et al. (2014)
Broccoli florets	300, 700 MPa/ 10 min/20, 82 °C	HPP can maintain a high level of intact glucosinolates in broccoli florets	Blok Frandsen et al. (2014)
Beetroot slices	650 MPa/ 3–30 min/RT	Great color instability, with high values for total color difference parameters (above 10)	Paciulli et al. (2016)

Flavor is one of the food properties most used by consumers to understand a product as consumable. Nevertheless, this kind of compounds is degraded by thermal processing, producing other compounds, such as volatile aldehydes, responsible for off-flavors development in vegetable products. Kebede et al. (2017) studied the effect of HPP (600 MPa/118 °C) on carrot purée and compared the formation of off-flavors compounds to a thermal-treated sample (Table 4.13). The authors reported a formation rate of that volatiles under high pressure ten times slower than at 0.1 MPa, indicating the importance of the processing conditions (temperature and time) on food quality (Kebede et al., 2017). Many similar results were reported by Kebede et al. (2014) after high-pressure/high-temperature treatment (600 MPa for 15 min at 117 °C) of onion, potato, pumpkin, and red beet.

In cruciferous vegetables, such as broccoli, glucosinolates are products responsible for their odor and taste. Some treatments of HPP can maintain a high level of intact glucosinolates in broccoli florets, mainly due to myrosinase inactivation at 700 MPa, 10 min, 20 °C, allowing preserving the quality of these vegetables (Blok Frandsen et al., 2014).

Table 4.14 Some publications with the main effects of HPP on the sensory properties of fish

Fish products	HPP conditions	Samples	Sensorial effects	Reference
Hake	200 and 400 MPa/ 5 min/7 °C	Samples cooked in boiling water for 5 min	General appearance and odor similar or slightly higher than control samples Higher flavor scores when compared to the control and a decrease with the pressure levels increase	Hurtado, Montero, and Borderias (2000)
Red mullet	220 and 330 MPa/ 5 and 3 min/25 °C	Raw fillets	No influence on appearance and odor	Erkan et al. (2010)
Sea bass	250 and 400 MPa/ 5 min/6 °C	Raw sea bass fillets	Color whitening (cooked fish appearance) Intensified brightness at 400 MPa (firmer fillets) No influence on the fresh odor High sensory acceptance	Teixeira et al. (2014)

4.2.3.2 Meat Products

Fish and Fish-Based Products

For consumer's perception, color is one of the most important sensory characteristics of fish muscles in determining their acceptability. For color measurements in fish processed by HPP is normally to use the CIE LAB system, obtaining values of $L*$ (lightness scale from 0 (black) to 100 (white)), $a*$ (scale ranging from −a (green) and +a (red)), and $b*$ (scale ranging from −b (blue) and +b (yellow)). From these parameters, it is possible to calculate the ΔE value, which is used to predict the differences in perception capacity ($\Delta E > 3.0$ = very distinctive, $1.5 < \Delta E < 3.0$ = distinct, and $\Delta E < 1.5$ = slightly distinct) (Adekunte, Tiwari, Cullen, Scannell, & O'Donnell, 2010). Normally, $L*$ value increases in HPP-treated fish, which shows more clear, gray, typical of cooked meat aspect after application of pressure levels between 150 and 300 MPa (Erkan, Üretener, & Alpas, 2010; Ramirez-Suarez & Morrissey, 2006; Sequeira-Munoz et al., 2006). Regarding $a*$ and $b*$ values, there are some differences in different published studies, but most of them have shown a decrease in $a*$ value (loss of red) (Erkan et al., 2010; Yagiz et al., 2007), and an increase in $b*$ value (up yellow) (Ramirez-Suarez & Morrissey, 2006; Sequeira-Munoz et al., 2006), which depends with the species and the pressure conditions.

The possible effects of HPP detected by instrumental methods are less noticeable during sensory evaluation. Few sensory studies assessed HPP-treated fish, but in general, the changes discretely influence the sensory attributes, most often positively (Table 4.14).

According to Hurtado et al. (2000), HPP-treated hake showed slightly higher scores for overall appearance and odor, as a function of pressure levels, compared to control samples. In addition, lower scores were observed within the pressure levels at 200 and 400 MPa. It is recognized that HPP increases whiteness index (instrumental measurement), giving opaque appearance typical of cooked product. Considering that hake is a white fish, whitening may have contributed to its better appearance. In the same way, Teixeira et al. (2014) verified that HPP caused the loss of red color on sea bass and increased whitening, which was correlated to color instrumental analyses. In addition, the same panel observed an increase in firmness with increasing pressure levels. The pressure levels did not influence fresh odor, although an increase in oxidation products was observed.

HPP has a significant effect on the elimination of microorganisms, inhibiting the production of biogenic amines, volatile nitrogen, and trimethylamine (Murchie et al., 2005), which contribute to improving the sensory quality, and ensure the safety of fish.

Meat and Meat-Based Products

Consumers request for fresh-like/high-quality food products have been increasing, and HPP as a novel non-thermal food processing technology has demonstrated to fulfill these requirements.

One of the most important attributes for consumers regarding meat products is its color. Shortly, it can be said that this parameter is related to the amount and chemical state of the hemoproteins and meat's structure, and its changes are related to ferrous myoglobin oxidation into ferric metmyoglobin. Although changes in several cured/processed meat products are acceptable by the consumers (depending on water content and water activity), HPP applied to fresh meat induces severe changes on this parameter (Sazonova et al., 2017).

Usually, color differences on fresh red meat occur when a pressure level above 200 MPa is applied over a few minutes at low temperatures. Briefly, the lightness ($L*$) parameter can change at a pressure level between 200 and 350 MPa (red into a paler pink), usually redness ($a*$) decreases resulting in a cooked-like appearance of the product at pressures of 400–500 MPa, and yellowness ($b*$) could increase or is not affected (Sazonova et al., 2017).

As myosin is sensitive to pressure it can denature between 180–300 MPa giving an opaque appearance similar to cooked meat. However, in the literature, it is stated that besides myosin unfolding on pressure treatment, some hydrogen bonded structure remains or is formed, being these destroyed by further heat treatment (Ma & Ledward, 2004; Ma & Ledward, 2013).

On the other hand, although myoglobin in pure solution at the same pH values found in meat is relatively stable to heat and pressure, on the product, the pigment is less stable to heat probably due to a pre-denaturation and conformational changes leading to more exposed/available haem to other denatured or denaturing proteins, so that it co-precipitates with them, being the final complex a brown denatured ferric haem pigment at normal pH (Ma & Ledward, 2013). The same is verified for pressure treatments, wherein solution myoglobin denaturation is reversible at all pH values except from those around its isoelectric point, however, in meat, it seems less stable than in solution, probably due to co-precipitation with other proteins and the complexes formed.

Some studies regarding color stability on high-pressure treatments concluded that this kind of technology could increase color stability over storage if a pressure treatment of 80–100 MPa after 2 days post-slaughter is applied. The same did not happen when the same pressure treatment was applied after 7 or 9 days post-slaughter, being this difference probably related to the destruction of the catalytic system responsible for the oxidation of myoglobin to metmyoglobin (Cheah & Ledward, 1997b).

Concerning meat flavor, it is not expected that HPP has a great impact on meat organoleptic characteristics, however as lipids and their oxidation products influence meat flavor and the enzymes responsible for the products formed over chill storage are susceptible to pressure, HPP can be responsible for meat flavor adulterations (Ma & Ledward, 2013).

By the years, it has been concluded that HPP turns polyunsaturated fatty acids in fresh meat more susceptible to oxidation, mainly above 400 MPa, being this impact detected by sensorial analyses (Cheah & Ledward, 1996; Ma, Ledward, Zamri, Frazier, & Zhou, 2007; Wang et al., 2013). This behavior has been explained by the release of metal ions (primarily iron) from the transition metal complexes present in meat, as well as by membrane disruption induced by pressure (Bolumar, Skibsted, & Orlien, 2012; Cheah & Ledward, 1997a; Orlien, Hansen, & Skibsted, 2000). It was also verified that supplementation with EDTA or vitamin E could lead to the lipid oxidation inhibition after pressure treatment, mainly for the former (Ma & Ledward, 2013). In Table 4.15, some works are listed regarding the HPP effect on the sensorial properties of meat-based products.

Table 4.15 Some studies regarding HPP effect on the sensorial properties of different meat-based products

Meat products	HPP conditions	Textural effects	Reference
Meat batters and reduced-fat sausages	200 MPa/ 2 min/10 °C	HPP sausages with 20% fat presented similar sensorial results when compared to AP sausages with 30% of fat	Yang et al. (2016)
Serrano dry-cured ham (30 different samples)	600 MPa/ 6 min/21 °C	Only 8 volatile compounds presented to be HPP affected (higher levels of methanethiol and sulfur dioxide in HPP-treated samples and higher levels of ethyl acetate, ethyl butanoate, ethyl 2-methylbutanoate, ethyl 3-methylbutanoate, dimethyl disulfide and dimethyl trisulfide in untreated samples)	Martínez-Onandi, Rivas-Cañedo, Nuñez, and Picon (2016)
"Lácon" (cured–cooked pork meat product)	500 and 600 MPa/ 5 min/≤16 °C	HPP did not show a considerable effect on the sensory characteristics of sliced "lacón", resulting in its stability over 120 days (at 4 °C), independently of HPP treatment	del Olmo, Calzada, and Nuñez (2014)
Ready-to-eat low-fat pastrami, Strassburg beef, export sausage, and Cajun beef	600 MPa/ 3 min/20 °C	Hedonic ratings revealed no difference in consumer acceptability between HPP and non-treated samples. There was no evidence of sensory quality deterioration over the study (98 days of refrigerated storage)	Hayman, Baxter, O'Riordan, and Stewart (2004)

Table 4.16 Examples of the main effects of HPP on the sensorial properties of milk

Dairy products	HPP conditions	Sensorial properties	Reference
Milk (bovine)	310 MPa/0.3 s	Continuous high-pressure throttling treatment changed milk color, producing lower *L*∗, *a*∗, and *b*∗ values	Adapa et al. (1997)
Milk (bovine)	300, 400, 500, and 676 MPa/5 min/4 °C	After HPP milk color became more yellow	Harte et al. (2003)
Milk (ovine)	100 to 500 MPa/ 10 and 30 min/25 and 50 °C	ΔE rates increased with pressure, with a maximum at 500 MPa, resulted from *L*∗ value reduction	Gervilla et al. (2001)

4.2.3.3 Dairy Products

Milk

Lightness value (*L*∗) was the main color parameter affected, contributing overall to the color losses in milk after HPP (Table 4.16). Gervilla et al. (2001) observed a decrease in *L*∗ and an increase ($p < 0.05$) of greenness (*a*∗) and yellowness (*b*∗) when the pressure increased. Similar results were obtained by Adapa et al. (1997), producing HPP darker (lower *L*∗ value), greener (lower *a*∗ value), and bluer (lower *b*∗ value) colors in milk. Casein micelles play an important role in light scattering and when HPP is applied, non-covalent forces (hydrogen bonds, ionic interactions, and hydrophobic forces) are disrupted, leading to casein micelles disintegration into small fragments that increase the translucence of the milk (Harte et al., 2003; Trujillo et al., 2007).

Fresh Cheese

After HPP, overall, fresh cheeses presented a more yellow color on the outside but without changes on the off-flavor parameters, when compared to non-processed ones (Table 4.17). Sandra et al. (2004) reported that HPP cheese was not significantly different for most attributes, however, HPP cheeses were slightly less crumbly than the control, but HPP cheeses were different from the control for all attributes, except color. In another study, HPP cheese obtained similar flavor scores to non-pressurized ones, with panelists familiar with fresh cheese more accepting of the texture than those not familiar fresh cheese (Van Hekken et al., 2013).

Panelists in the Evert-Arriagada et al. (2012) study, identified the pressurized fresh cheeses (300–400 MPa, 5 min) as more yellow, firmer, and less watery, but without off-flavors or great differences in flavor and aroma. When a higher pressure was applied (500 MPa for 5 min), an increase in firmness in high pressure-treated cheeses was noticeable, while flavor, aroma, elasticity and off-flavor parameters remained unchanged with HPP treatment of fresh cheese did not affect panelists' preference (Evert-Arriagada et al., 2014).

Table 4.17 Examples of the main effects of HPP on the sensorial properties of fresh cheese

Dairy products	HPP conditions	Sensorial properties	Reference
Fresh Cheese (cow)	600 MPa/3 or 10 min/20 °C	Panelists were able to distinguish the control from HPP cheeses, scoring around 3.4 "moderately liked"	Van Hekken et al. (2013)
Fresh Cheese (cow)	500 MPa/5 min/16 °C	Panelist equally preferred pressurized to non-treated cheeses, although HPP increased firmness in high pressure-treated cheeses	Evert-Arriagada et al. (2014)
Fresh Cheese (cow)	400 MPa/20 min/20 °C	HP cheese was slightly less crumbly than the control Stickiness and sticky residuals of HP cheese increased during storage	Sandra et al. (2004)
Fresh Cheese (cow)	300 or 400 MPa/5 min/6 °C	Panelists classified pressurized cheeses as more yellow, firmer, and less watery, but without off-flavors or great differences in flavor and aroma when compared with the reference cheeses	Evert-Arriagada et al. (2014)

Table 4.18 Effect of HPP on the sensorial properties of different cheeses

Dairy products	HPP conditions	Sensorial effects	Reference
Brie cheese (pasteurized cow milk)	400 or 600 MPa/5 min/9–14 °C	Flavor quality, intensity, and bitterness: did not vary immediately after HPP (at 21 days of ripening)	Calzada, del Olmo, Picon, Gaya, et al. (2014c)
Blue-veined cheese (pasteurized ewe milk)	400 or 600 MPa/5 min/13 °C	Flavor intensity and quality: scores of HPP cheeses did not differ from those of control cheese (exception of Cheeses HPP at 600 MPa at 21 days of ripening, with lower scores)	Calzada, Del Olmo, Picon, Gaya, and Nuñez (2013)
NA (raw cow milk)	400 or 600 MPa/5 min/14 °C	Odor intensity: did not differ significantly among HPP and control cheeses	Calzada et al. (2015)
Ibores Cheese (raw goat milk)	400 or 600 MPa/7 min/10 °C	Odor intensity, flavor intensity, and taste: higher scores for HPP cheeses (better when were HPP at 30 days of ripening)	Delgado et al. (2012)
Casar Cheese (raw ewe milk)	400 or 600 MPa/5 min/14 °C	Flavor intensity and quality: similar scores of HPP cheeses and control cheese (at 60 days of ripening); with time HPP win flavor quality	Calzada, del Olmo, Picon, Gaya, and Nuñez (2014a)

Cheese

In ripened cheese, interesting sensorial results were obtained after cheese HPP. In general, immediately after treatment, the flavor quality and intensity scores were closer to control cheese (Calzada et al., 2015; Calzada, del Olmo, Picon, Gaya, et al., 2014a; Calzada, del Olmo, Picon, & Nuñez, 2014c), but after some days of storage, an increase of these sensorial attributes were verified (Delgado et al., 2012) (Table 4.18).

4.3 Conclusion

The consumers' demand for new and healthier food products with distinctive organoleptic characteristics led to the application of novel food technologies to fulfill their requests. Currently, HPP is one of such novel technologies and has been applied, for instance, on the production of fruit juices and ready-to-eat meals, where higher nutritional foods with exceptional sensory quality are attained. This non-thermal food processing technology not only is capable to guarantee food safety, increasing food products shelf lives and the reduction/avoidance of chemical preservatives, but also can result in fresher-tasting foods. Along with this chapter, the HPP impact on different foods such as fruits, vegetables, meat, and dairy products, was revised and discussed. In the end, it was possible to perceive that HPP, when applied at optimal conditions for a specific product, allows obtaining several benefits depending on the product. It is expected that research regarding HPP effect on foods will continue since its application for new food products' development has been increasing due to novel organoleptic characteristics' achievement, for instance, texture properties' modification. Thus, it is very likely that in a few years the knowledge concerning HPP of different foods not yet studied and, on its constituents, will increase, as well HPP combined to other food technologies could be a reality.

References

Adapa, S., Schmidt, K. A., & Toledo, R. (1997). Functional properties of skim milk processed with continuous high pressure throttling. *Journal of Dairy Science, 80*(9), 1941–1948.

Adekunte, A. O., Tiwari, B. K., Cullen, P. J., Scannell, A. G. M., & O'Donnell, C. P. (2010). Effect of sonication on colour, ascorbic acid and yeast inactivation in tomato juice. *Food Chemistry, 122*(3), 500–507.

Amanatidou, A., Schlüter, O., Lemkau, K., Gorris, L. G. M., Smid, E. J., & Knorr, D. (2000). Effect of combined application of high pressure treatment and modified atmospheres on the shelf life of fresh Atlantic salmon. *Innovative Food Science & Emerging Technologies, 1*(2), 87–98.

Angsupanich, K., & Ledward, D. A. (1998). High pressure treatment effects on cod (*Gadus morhua*) muscle. *Food Chemistry, 63*(1), 39–50.

Arnold, C., Schwarzenbolz, U., & Böhm, V. (2014). Carotenoids and chlorophylls in processed xanthophyll-rich food. *LWT – Food Science and Technology, 57*(1), 442–445.

Aubourg, S. P., Torres, J. A., Saraiva, J. A., Guerra-Rodríguez, E., & Vázquez, M. (2013). Effect of high-pressure treatments applied before freezing and frozen storage on the functional and sensory properties of Atlantic mackerel (*Scomber scombrus*). *LWT – Food Science and Technology, 53*(1), 100–106.

Barrett, D. M., & Lloyd, B. (2012). Advanced preservation methods and nutrient retention in fruits and vegetables. *Journal of the Science of Food and Agriculture, 92*(1), 7–22.

Basak, S., & Ramaswamy, H. S. (1998). Effect of high pressure processing on the texture of selected fruits and vegetables. *Journal of Texture Studies, 29*(5), 587–601.

Baxter, I. A., Easton, K., Schneebeli, K., & Whitfield, F. B. (2005). High pressure processing of Australian navel orange juices: Sensory analysis and volatile flavor profiling. *Innovative Food Science & Emerging Technologies, 6*(4), 372–387.

Beilken, S. L., Macfarlane, J. J., & Jones, P. N. (1990). Effect of high pressure during heat treatment on the wamer-bratzler shear force values of selected beef muscles. *Journal of Food Science, 55*(1), 15–18.

Blok Frandsen, H., Ejdrup Markedal, K., Martín-Belloso, O., Sánchez-Vega, R., Soliva-Fortuny, R., Sørensen, H., … Sørensen, J. C. (2014). Effects of novel processing techniques on glucosinolates and membrane associated myrosinases in broccoli. *Polish Journal of Food and Nutrition Sciences, 64*(1), 17–25.

Bolumar, T., Skibsted, L. H., & Orlien, V. (2012). Kinetics of the formation of radicals in meat during high pressure processing. *Food Chemistry, 134*(4), 2114–2120.

Calzada, J., Del Olmo, A., Picon, A., Gaya, P., & Nuñez, M. (2013). Proteolysis and biogenic amine buildup in high-pressure treated ovine milk blue-veined cheese. *Journal of Dairy Science, 96*(8), 4816–4829.

Calzada, J., del Olmo, A., Picon, A., & Nuñez, M. (2015). Effect of high pressure processing on the lipolysis, volatile compounds, odour and colour of cheese made from unpasteurized milk. *Food and Bioprocess Technology, 8*(5), 1076–1088.

Calzada, J., del Olmo, A., Picon, A., Gaya, P., & Nuñez, M. (2014a). Using high-pressure processing for reduction of proteolysis and prevention of over-ripening of raw milk cheese. *Food and Bioprocess Technology, 7*(5), 1404–1413.

Calzada, J., del Olmo, A., Picon, A., Gaya, P., & Nuñez, M. (2014b). Effect of high-pressure-processing on the microbiology, proteolysis, texture and flavour of Brie cheese during ripening and refrigerated storage. *International Dairy Journal, 37*(2), 64–73.

Calzada, J., del Olmo, A., Picon, A., & Nuñez, M. (2014c). Effect of high-pressure-processing on lipolysis and volatile compounds of Brie cheese during ripening and refrigerated storage. *International Dairy Journal, 39*(2), 232–239.

Capellas, M., Mor-Mur, M., Sendra, E., & Guamis, B. (2001). Effect of high-pressure processing on physico-chemical characteristics of fresh goats' milk cheese (Mató). *International Dairy Journal, 11*(3), 165–173.

Chang, Y.-H., Wu, S.-J., Chen, B.-Y., Huang, H.-W., & Wang, C.-Y. (2017). Effect of high-pressure processing and thermal pasteurization on overall quality parameters of white grape juice. *Journal of the Science of Food and Agriculture, 97*(10), 3166–3172.

Cheah, P. B., & Ledward, D. A. (1996). High pressure effects on lipid oxidation in minced pork. *Meat Science, 43*(2), 123–134.

Cheah, P. B., & Ledward, D. A. (1997a). Catalytic mechanism of lipid oxidation following high pressure treatment in pork fat and meat. *Journal of Food Science, 62*(6), 1135–1139.

Cheah, P. B., & Ledward, D. A. (1997b). Inhibition of metmyoglobin formation in fresh beef by pressure treatment. *Meat Science, 45*(3), 411–418.

Chen, D., Xi, H., Guo, X., Qin, Z., Pang, X., Hu, X., … Wu, J. (2013). Comparative study of quality of cloudy pomegranate juice treated by high hydrostatic pressure and high temperature short time. *Innovative Food Science & Emerging Technologies, 19*, 85–94.

Chéret, R., Chapleau, N., Delbarre-Ladrat, C., Verrez-Bagnis, V., & de Lamballerie, M. (2005). Effects of high pressure on texture and microstructure of sea bass (*Dicentrarchus labrax* L.). *Journal of Food Science, 70*(8), e477–e483.

Chevalier, D., Le Bail, A., & Ghoul, M. (2001). Effects of high pressure treatment (100–200 MPa) at low temperature on turbot (*Scophthalmus maximus*) muscle. *Food Research International, 34*(5), 425–429.

Clariana, M., Guerrero, L., Sárraga, C., Díaz, I., Valero, Á., & García-Regueiro, J. A. (2011). Influence of high pressure application on the nutritional, sensory and microbiological characteristics of sliced skin vacuum packed dry-cured ham. Effects along the storage period. *Innovative Food Science & Emerging Technologies, 12*(4), 456–465.

Contador, R., González-Cebrino, F., García-Parra, J., Lozano, M., & Ramírez, R. (2014). Effect of hydrostatic high pressure and thermal treatments on two types of pumpkin purée and changes during refrigerated storage. *Journal of Food Processing and Preservation, 38*(2), 704–712.

de Oliveira, M. M., Tribst, A. A. L., Leite Júnior, B. R. D. C., de Oliveira, R. A., & Cristianini, M. (2015). Effects of high pressure processing on cocoyam, Peruvian carrot, and sweet potato: Changes in microstructure, physical characteristics, starch, and drying rate. *Innovative Food Science & Emerging Technologies, 31*, 45–53.

del Olmo, A., Calzada, J., & Nuñez, M. (2014). Effect of high pressure processing and modified atmosphere packaging on the safety and quality of sliced ready-to-eat "lacón", a cured–cooked pork meat product. *Innovative Food Science & Emerging Technologies, 23*, 25–32.

Delgado, F. J., González-Crespo, J., Cava, R., & Ramírez, R. (2012). Changes in microbiology, proteolysis, texture and sensory characteristics of raw goat milk cheeses treated by high-pressure at different stages of maturation. *LWT – Food Science and Technology, 48*(2), 268–275.

Delgado, F. J., Rodríguez-Pinilla, J., Márquez, G., Roa, I., & Ramírez, R. (2015). Physicochemical, proteolysis and texture changes during the storage of a mature soft cheese treated by high-pressure hydrostatic. *European Food Research and Technology, 240*(6), 1167–1176.

Denoya, G. I., Vaudagna, S. R., & Polenta, G. (2015). Effect of high pressure processing and vacuum packaging on the preservation of fresh-cut peaches. *LWT – Food Science and Technology, 62*(1, Part 2), 801–806.

Dhineshkumar, V., Ramasamy, D., & Siddharth, M. (2016). High pressure processing technology in dairy processing: A review. *Asian Journal of Dairy and Food Research, 35*(2), 87–95.

Dörnenburg, H., & Knorr, D. (1998). Monitoring the impact of high-pressure processing on the biosynthesis of plant metabolites using plant cell cultures. *Trends in Food Science & Technology, 9*(10), 355–361.

Erkan, N., Üretener, G., & Alpas, H. (2010). Effect of high pressure (HP) on the quality and shelf life of red mullet (*Mullus surmelutus*). *Innovative Food Science & Emerging Technologies, 11*(2), 259–264.

Evert-Arriagada, K., Hernández-Herrero, M. M., Guamis, B., & Trujillo, A. J. (2014). Commercial application of high-pressure processing for increasing starter-free fresh cheese shelf-life. *LWT – Food Science and Technology, 55*(2), 498–505.

Evert-Arriagada, K., Hernández-Herrero, M. M., Juan, B., Guamis, B., & Trujillo, A. J. (2012). Effect of high pressure on fresh cheese shelf-life. *Journal of Food Engineering, 110*(2), 248–253.

Figueirêdo, B. C., Bragagnolo, N., Skibsted, L. H., & Orlien, V. (2015). Inhibition of cholesterol and polyunsaturated fatty acids oxidation through the use of annatto and bixin in high-pressure processed fish. *Journal of Food Science, 80*(8), C1646–C1653.

Gao, G., Ren, P., Cao, X., Yan, B., Liao, X., Sun, Z., & Wang, Y. (2016). Comparing quality changes of cupped strawberry treated by high hydrostatic pressure and thermal processing during storage. *Food and Bioproducts Processing, 100*, 221–229.

García-Parra, J., González-Cebrino, F., Delgado, J., Cava, R., & Ramírez, R. (2016). High pressure assisted thermal processing of pumpkin purée: Effect on microbial counts, color, bioactive compounds and polyphenoloxidase enzyme. *Food and Bioproducts Processing, 98*, 124–132.

Garde, S., Arqués, J. L., Gaya, P., Medina, M., & Nuñez, M. (2007). Effect of high-pressure treatments on proteolysis and texture of ewes' raw milk La Serena cheese. *International Dairy Journal, 17*(12), 1424–1433.

Gervilla, R., Ferragut, V., & Guamis, B. (2001). High hydrostatic pressure effects on color and milk-fat globule of ewe's milk. *Journal of Food Science, 66*(6), 880–885.

Grossi, A., Søltoft-Jensen, J., Knudsen, J. C., Christensen, M., & Orlien, V. (2011). Synergistic cooperation of high pressure and carrot dietary fibre on texture and colour of pork sausages. *Meat Science, 89*(2), 195–201.

Gupta, R., Kopec, R. E., Schwartz, S. J., & Balasubramaniam, V. M. (2011). Combined pressure-temperature effects on carotenoid retention and bioaccessibility in tomato juice. *Journal of Agricultural and Food Chemistry, 59*(14), 7808–7817.

Guyon, C., Meynier, A., & de Lamballerie, M. (2016). Protein and lipid oxidation in meat: A review with emphasis on high-pressure treatments. *Trends in Food Science & Technology, 50*, 131–143.

Harte, F., Luedecke, L., Swanson, B., & Barbosa-Cánovas, G. V. (2003). Low-fat set yogurt made from milk subjected to combinations of high hydrostatic pressure and thermal processing. *Journal of Dairy Science, 86*(4), 1074–1082.

Hartyáni, P., Dalmadi, I., Cserhalmi, Z., Kántor, D.-B., Tóth-Markus, M., & Sass-Kiss, Á. (2011). Physical–chemical and sensory properties of pulsed electric field and high hydrostatic pressure treated citrus juices. *Innovative Food Science & Emerging Technologies, 12*(3), 255–260.

Hayman, M. M., Baxter, I., O'Riordan, P. J., & Stewart, C. M. (2004). Effects of high-pressure processing on the safety, quality, and shelf life of ready-to-eat meats. *Journal of Food Protection, 67*(8), 1709–1718.

He, Z., Huang, Y., Li, H., Qin, G., Wang, T., & Yang, J. (2012). Effect of high-pressure treatment on the fatty acid composition of intramuscular lipid in pork. *Meat Science, 90*(1), 170–175.

Hernández-Carrión, M., Hernando, I., & Quiles, A. (2014). High hydrostatic pressure treatment as an alternative to pasteurization to maintain bioactive compound content and texture in red sweet pepper. *Innovative Food Science & Emerging Technologies, 26*, 76–85.

Homma, N., Ikeuchi, Y., & Suzuki, A. (1994). Effects of high pressure treatment on the proteolytic enzymes in meat. *Meat Science, 38*(2), 219–228.

Huang, H.-W., Wu, S.-J., Lu, J.-K., Shyu, Y.-T., & Wang, C.-Y. (2017). Current status and future trends of high-pressure processing in food industry. *Food Control, 72*, 1–8.

Huppertz, T., Kelly, A. L., & Fox, P. F. (2002). Effects of high pressure on constituents and properties of milk. *International Dairy Journal, 12*(7), 561–572.

Hurtado, A., Guàrdia, M. D., Picouet, P., Jofré, A., Ros, J. M., & Bañón, S. (2017). Stabilisation of red fruit-based smoothies by high-pressure processing. Part II: Effects on sensory quality and selected nutrients. *Journal of the Science of Food and Agriculture, 97*(3), 777–783.

Hurtado, J. L., Montero, P., & Borderias, A. J. (2000). Extension of shelf life of chilled hake (*Merluccius capensis*) by high pressure/Prolongacion de la vida util de merluza (*Merluccius capensis*) sometida a altas presiones conservada en refrigeracion. *Food Science and Technology International, 6*(3), 243–249.

Jacobo-Velázquez, D. A., Cuéllar-Villarreal, M. d. R., Welti-Chanes, J., Cisneros-Zevallos, L., Ramos-Parra, P. A., & Hernández-Brenes, C. (2017). Nonthermal processing technologies as elicitors to induce the biosynthesis and accumulation of nutraceuticals in plant foods. *Trends in Food Science & Technology, 60*, 80–87.

Juan, B., Ferragut, V., Guamis, B., & Trujillo, A.-J. (2008). The effect of high-pressure treatment at 300MPa on ripening of ewes' milk cheese. *International Dairy Journal, 18*(2), 129–138.

Jung, L.-S., Lee, S. H., Kim, S., & Ahn, J. (2013). Effect of high hydrostatic pressure on the quality-related properties of carrot and spinach. *Food Science and Biotechnology, 22*(1), 189–195.

Kaur, L., Astruc, T., Vénien, A., Loison, O., Cui, J., Irastorza, M., & Boland, M. (2016). High pressure processing of meat: Effects on ultrastructure and protein digestibility. *Food & Function, 7*(5), 2389–2397.

Kebede, B., Grauwet, T., Andargie, T., Sempiri, G., Palmers, S., Hendrickx, M., & Van Loey, A. (2017). Kinetics of Strecker aldehyde formation during thermal and high pressure high temperature processing of carrot puree. *Innovative Food Science & Emerging Technologies, 39*, 88–93.

Kebede, B. T., Grauwet, T., Mutsokoti, L., Palmers, S., Vervoort, L., Hendrickx, M., & Van Loey, A. (2014). Comparing the impact of high pressure high temperature and thermal sterilization on the volatile fingerprint of onion, potato, pumpkin and red beet. *Food Research International, 56*, 218–225.

Koutchma, T., Popović, V., Ros-Polski, V., & Popielarz, A. (2016). Effects of ultraviolet light and high-pressure processing on quality and health-related constituents of fresh juice products. *Comprehensive Reviews in Food Science and Food Safety, 15*(5), 844–867.

Lakshmanan, R., Patterson, M. F., & Piggott, J. R. (2005). Effects of high-pressure processing on proteolytic enzymes and proteins in cold-smoked salmon during refrigerated storage. *Food Chemistry, 90*(4), 541–548.

Landl, A., Abadias, M., Sárraga, C., Viñas, I., & Picouet, P. A. (2010). Effect of high pressure processing on the quality of acidified Granny Smith apple purée product. *Innovative Food Science & Emerging Technologies, 11*(4), 557–564.

Liu, F., Zhang, X., Zhao, L., Wang, Y., & Liao, X. (2016). Potential of high-pressure processing and high-temperature/short-time thermal processing on microbial, physicochemical and sensory assurance of clear cucumber juice. *Innovative Food Science & Emerging Technologies, 34*, 51–58.

López-Fandiño, R., Fuente, M. A. D. L., Ramos, M., & Olano, A. (1998). Distribution of minerals and proteins between the soluble and colloidal phases of pressurized milks from different species. *Journal of Dairy Research, 65*(1), 69–78.

Ma, H., & Ledward, D. A. (2013). High pressure processing of fresh meat - is it worth it? *Meat Science, 95*(4), 897–903.

Ma, H.-J., & Ledward, D. A. (2004). High pressure/thermal treatment effects on the texture of beef muscle. *Meat Science, 68*(3), 347–355.

Ma, H. J., Ledward, D. A., Zamri, A. I., Frazier, R. A., & Zhou, G. H. (2007). Effects of high pressure/thermal treatment on lipid oxidation in beef and chicken muscle. *Food Chemistry, 104*(4), 1575–1579.

Macfarlane, J. J. (1973). Pre-rigor pressurization of muscle: Effects on pH, shear value and taste panel assessment. *Journal of Food Science, 38*(2), 294–298.

Marcos, B., Aymerich, T., Dolors Guardia, M., & Garriga, M. (2007). Assessment of high hydrostatic pressure and starter culture on the quality properties of low-acid fermented sausages. *Meat Science, 76*(1), 46–53.

Marcos, B., Kerry, J. P., & Mullen, A. M. (2010). High pressure induced changes on sarcoplasmic protein fraction and quality indicators. *Meat Science, 85*(1), 115–120.

Marszałek, K., Woźniak, Ł., Kruszewski, B., & Skąpska, S. (2017). The Effect of high pressure techniques on the stability of anthocyanins in fruit and vegetables. *International Journal of Molecular Sciences, 18*(2). https://doi.org/10.3390/ijms18020277

Martínez-Onandi, N., Castioni, A., San Martín, E., Rivas-Cañedo, A., Nuñez, M., Torriani, S., & Picon, A. (2017). Microbiota of high-pressure-processed Serrano ham investigated by culture-dependent and culture-independent methods. *International Journal of Food Microbiology, 241*, 298–307.

Martínez-Onandi, N., Rivas-Cañedo, A., Nuñez, M., & Picon, A. (2016). Effect of chemical composition and high pressure processing on the volatile fraction of Serrano dry-cured ham. *Meat Science, 111*, 130–138.

Martínez-Rodríguez, Y., Acosta-Muñiz, C., Olivas, G. I., Guerrero-Beltrán, J., Rodrigo-Aliaga, D., & Sepúlveda, D. R. (2012). High hydrostatic pressure processing of cheese. *Comprehensive Reviews in Food Science and Food Safety, 11*(4), 399–416.

McArdle, R., Marcos, B., Kerry, J. P., & Mullen, A. (2010). Monitoring the effects of high pressure processing and temperature on selected beef quality attributes. *Meat Science, 86*(3), 629–634.

Medina-Meza, I. G., Barnaba, C., & Barbosa-Cánovas, G. V. (2014). Effects of high pressure processing on lipid oxidation: A review. *Innovative Food Science & Emerging Technologies, 22*, 1–10.

Miguel-Pintado, C., Nogales, S., Fernández-León, A. M., Delgado-Adámez, J., Hernández, T., Lozano, M., … Ramírez, R. (2013). Effect of hydrostatic high pressure processing on nectarine halves pretreated with ascorbic acid and calcium during refrigerated storage. *LWT – Food Science and Technology, 54*(1), 278–284.

Moltó-Puigmartí, C., Permanyer, M., Castellote, A. I., & López-Sabater, M. C. (2011). Effects of pasteurisation and high-pressure processing on vitamin C, tocopherols and fatty acids in mature human milk. *Food Chemistry, 124*(3), 697–702.

Montiel, R., De Alba, M., Bravo, D., Gaya, P., & Medina, M. (2012). Effect of high pressure treatments on smoked cod quality during refrigerated storage. *Food Control, 23*(2), 429–436. https://doi.org/10.1016/j.foodcont.2011.08.011

Morton, J. D., Pearson, R. G., Lee, H. Y.-Y., Smithson, S., Mason, S. L., & Bickerstaffe, R. (2017). High pressure processing improves the tenderness and quality of hot-boned beef. *Meat Science, 133*, 69–74.

Murchie, L. W., Cruz-Romero, M., Kerry, J. P., Linton, M., Patterson, M. F., Smiddy, M., & Kelly, A. L. (2005). High pressure processing of shellfish: A review of microbiological and other quality aspects. *Innovative Food Science & Emerging Technologies, 6*(3), 257–270.

O'Reilly, C. E., Kelly, A. L., Murphy, P. M., & Beresford, T. P. (2001). High pressure treatment: Applications in cheese manufacture and ripening. *Trends in Food Science & Technology, 12*(2), 51–59.

Oey, I., Van der Plancken, I., Van Loey, A., & Hendrickx, M. (2008). Does high pressure processing influence nutritional aspects of plant based food systems? *Trends in Food Science & Technology, 19*(6), 300–308.

Okpala, C. O. R., Piggott, J. R., & Schaschke, C. J. (2010). Influence of high-pressure processing (HPP) on physico-chemical properties of fresh cheese. *Innovative Food Science & Emerging Technologies, 11*(1), 61–67.

Orlien, V., Hansen, E., & Skibsted, L. H. (2000). Lipid oxidation in high-pressure processed chicken breast muscle during chill storage: Critical working pressure in relation to oxidation mechanism. *European Food Research and Technology, 211*(2), 99–104.

Ortea, I., Rodríguez, A., Tabilo-Munizaga, G., Pérez-Won, M., & Aubourg, S. P. (2010). Effect of hydrostatic high-pressure treatment on proteins, lipids and nucleotides in chilled farmed salmon (*Oncorhynchus kisutch*) muscle. *European Food Research and Technology, 230*(6), 925–934.

Paciulli, M., Medina-Meza, I. G., Chiavaro, E., & Barbosa-Cánovas, G. V. (2016). Impact of thermal and high pressure processing on quality parameters of beetroot (*Beta vulgaris* L.). *LWT – Food Science and Technology, 68*, 98–104.

Pasha, I., Saeed, F., Sultan, M. T., Khan, M. R., & Rohi, M. (2014). Recent developments in minimal processing: A tool to retain nutritional quality of food. *Critical Reviews in Food Science and Nutrition, 54*(3), 340–351.

Patras, A., Brunton, N., Da Pieve, S., Butler, F., & Downey, G. (2009). Effect of thermal and high pressure processing on antioxidant activity and instrumental colour of tomato and carrot purées. *Innovative Food Science & Emerging Technologies, 10*(1), 16–22.

Patras, A., Brunton, N. P., Da Pieve, S., & Butler, F. (2009). Impact of high pressure processing on total antioxidant activity, phenolic, ascorbic acid, anthocyanin content and colour of strawberry and blackberry purées. *Innovative Food Science & Emerging Technologies, 10*(3), 308–313.

Picouet, P. A., Hurtado, A., Jofré, A., Bañon, S., Ros, J.-M., & Guàrdia, M. D. (2016). Effects of thermal and high-pressure treatments on the microbiological, nutritional and sensory quality of a multi-fruit smoothie. *Food and Bioprocess Technology, 9*(7), 1219–1232.

Plaza, L., Sánchez-Moreno, C., De Ancos, B., Elez-Martínez, P., Martín-Belloso, O., & Cano, M. P. (2011). Carotenoid and flavanone content during refrigerated storage of orange juice processed by high-pressure, pulsed electric fields and low pasteurization. *LWT – Food Science and Technology, 44*(4), 834–839.

Ramirez-Suarez, J. C., & Morrissey, M. T. (2006). Effect of high pressure processing (HPP) on shelf life of albacore tuna (*Thunnus alalunga*) minced muscle. *Innovative Food Science & Emerging Technologies, 7*(1), 19–27.

Rivas-Cañedo, A., Fernández-García, E., & Nuñez, M. (2009). Volatile compounds in fresh meats subjected to high pressure processing: Effect of the packaging material. *Meat Science, 81*(2), 321–328.

Rux, G., Schlüter, O., Geyer, M., & Herppich, W. B. (2017). Characterization of high hydrostatic pressure effects on fresh produce cell turgor using pressure probe analyses. *Postharvest Biology and Technology, 132*, 188–194.

Ruxton, C. H. S., Calder, P. C., Reed, S. C., & Simpson, M. J. A. (2005). The impact of long-chain n-3 polyunsaturated fatty acids on human health. *Nutrition Research Reviews, 18*(1), 113–129.

Saldo, J., Fernández, A., Sendra, E., Butz, P., Tauscher, B., & Guamis, B. (2003). High pressure treatment decelerates the lipolysis in a caprine cheese. *Food Research International, 36*(9), 1061–1068.

Saldo, J., McSweeney, P. L. H., Sendra, E., Kelly, A. L., & Guamis, B. (2002). Proteolysis in caprine milk cheese treated by high pressure to accelerate cheese ripening. *International Dairy Journal, 12*(1), 35–44.

Sánchez, C., Baranda, A. B., & Martínez de Marañón, I. (2014). The effect of high pressure and high temperature processing on carotenoids and chlorophylls content in some vegetables. *Food Chemistry, 163*, 37–45.

Sánchez-Moreno, C., de Ancos, B., Plaza, L., Elez-Martínez, P., & Cano, M. P. (2009). Nutritional approaches and health-related properties of plant foods processed by high pressure and pulsed electric fields. *Critical Reviews in Food Science and Nutrition, 49*(6), 552–576.

Sánchez-Moreno, C., Plaza, L., De Ancos, B., & Cano, M. P. (2003). Vitamin C, provitamin A carotenoids, and other carotenoids in high-pressurized orange juice during refrigerated storage. *Journal of Agricultural and Food Chemistry, 51*(3), 647–653.

Sánchez-Moreno, C., Plaza, L., Elez-Martínez, P., De Ancos, B., Martín-Belloso, O., & Cano, M. P. (2005). Impact of high pressure and pulsed electric fields on bioactive compounds and antioxidant activity of orange juice in comparison with traditional thermal processing. *Journal of Agricultural and Food Chemistry, 53*(11), 4403–4409.

Sandra, S., Stanford, M. A., & Goddik, L. M. (2004). The use of high-pressure processing in the production of Queso Fresco cheese. *Journal of Food Science, 69*(4), FEP153–FEP158.

Sazonova, S., Galoburda, R., & Gramatina, I. (2017). Application of high-pressure processing for safety and shelf-life quality of meat – a review. Presented at the Baltic Conference on Food Science and Technology FOODBALT "Food for consumer well-being." https://doi.org/10.22616/FoodBalt.2017.001

Sequeira-Munoz, A., Chevalier, D., LeBail, A., Ramaswamy, H. S., & Simpson, B. K. (2006). Physicochemical changes induced in carp (*Cyprinus carpio*) fillets by high pressure processing at low temperature. *Innovative Food Science & Emerging Technologies, 7*(1), 13–18.

Sikes, A., Tornberg, E., & Tume, R. (2010). A proposed mechanism of tenderising post-rigor beef using high pressure–heat treatment. *Meat Science, 84*(3), 390–399.

Sikes, A. L., & Warner, R. (2016). Application of high hydrostatic pressure for meat tenderization. In K. Knoerzer, P. Juliano, & G. Smithers (Eds.), *Innovative food processing technologies: Extraction, separation, component modification and process intensification* (pp. 259–290), Woodhead Publishing, Cambridge.

Souza, C. M., Boler, D. D., Clark, D. L., Kutzler, L. W., Holmer, S. F., Summerfield, J. W., … Killefer, J. (2011). The effects of high pressure processing on pork quality, palatability, and further processed products. *Meat Science, 87*(4), 419–427.

Sun, X. D., & Holley, R. A. (2010). High hydrostatic pressure effects on the texture of meat and meat products. *Journal of Food Science, 75*(1), R17–R23.

Tamm, A., Bolumar, T., Bajovic, B., & Toepfl, S. (2016). Salt (NaCl) reduction in cooked ham by a combined approach of high pressure treatment and the salt replacer KCl. *Innovative Food Science & Emerging Technologies, 36*, 294–302.

Teixeira, B., Fidalgo, L., Mendes, R., Costa, G., Cordeiro, C., Marques, A., … Nunes, M. L. (2013). Changes of enzymes activity and protein profiles caused by high-pressure processing in sea bass (*Dicentrarchus labrax*) fillets. *Journal of Agricultural and Food Chemistry, 61*(11), 2851–2860.

Teixeira, B., Marques, A., Mendes, R., Gonçalves, A., Fidalgo, L., Oliveira, M., … Nunes, M. L. (2014). Effects of high-pressure processing on the quality of sea bass (*Dicentrarchus labrax*) fillets during refrigerated storage. *Food and Bioprocess Technology, 7*(5), 1333–1343.

Tewari, S., Sehrawat, R., Nema, P. K., & Kaur, B. P. (2017). Preservation effect of high pressure processing on ascorbic acid of fruits and vegetables: A review. *Journal of Food Biochemistry, 41*(1), e12319. https://doi.org/10.1111/jfbc.12319

Torres, B., Tiwari, B. K., Patras, A., Cullen, P. J., Brunton, N., & O'Donnell, C. P. (2011). Stability of anthocyanins and ascorbic acid of high pressure processed blood orange juice during storage. *Innovative Food Science & Emerging Technologies, 12*(2), 93–97.

Trejo Araya, X. I., Hendrickx, M., Verlinden, B. E., Van Buggenhout, S., Smale, N. J., Stewart, C., & John Mawson, A. (2007). Understanding texture changes of high pressure processed fresh carrots: A microstructural and biochemical approach. *Journal of Food Engineering, 80*(3), 873–884.

Tribst, A. A. L., Leite Júnior, B. R. D. C., de Oliveira, M. M., & Cristianini, M. (2016). High pressure processing of cocoyam, Peruvian carrot and sweet potato: Effect on oxidative enzymes and impact in the tuber color. *Innovative Food Science & Emerging Technologies, 34*, 302–309.

Trujillo, A. J., Castro, N., Quevedo, J. M., Argüello, A., Capote, J., & Guamis, B. (2007). Effect of heat and high-pressure treatments on microbiological quality and immunoglobulin G stability of caprine colostrum. *Journal of Dairy Science, 90*(2), 833–839.

Trujillo, A. J., Capellas, M., Saldo, J., Gervilla, R., & Guamis, B. (2002). Applications of high-hydrostatic pressure on milk and dairy products: A review. *Innovative Food Science & Emerging Technologies, 3*(4), 295–307.

Valdramidis, V. P., Graham, W. D., Beattie, A., Linton, M., McKay, A., Fearon, A. M., & Patterson, M. F. (2009). Defining the stability interfaces of apple juice: Implications on the optimisation and design of High Hydrostatic Pressure treatment. *Innovative Food Science & Emerging Technologies, 10*(4), 396–404.

Van Hekken, D. L., Tunick, M. H., Farkye, N. Y., & Tomasula, P. M. (2013). Effect of hydrostatic high-pressure processing on the chemical, functional, and rheological properties of starter-free Queso Fresco. *Journal of Dairy Science, 96*(10), 6147–6160.

Vázquez-Gutiérrez, J. L., Quiles, A., Hernando, I., & Pérez-Munuera, I. (2011). Changes in the microstructure and location of some bioactive compounds in persimmons treated by high hydrostatic pressure. *Postharvest Biology and Technology, 61*(2), 137–144.

Voigt, D. D., Chevalier, F., Qian, M. C., & Kelly, A. L. (2010). Effect of high-pressure treatment on microbiology, proteolysis, lipolysis and levels of flavour compounds in mature blue-veined cheese. *Innovative Food Science & Emerging Technologies, 11*(1), 68–77.

Wang, C.-Y., Huang, H.-W., Hsu, C.-P., & Yang, B. B. (2016). Recent advances in food processing using high hydrostatic pressure technology. *Critical Reviews in Food Science and Nutrition, 56*(4), 527–540.

Wang, Q., Zhao, X., Ren, Y., Fan, E., Chang, H., & Wu, H. (2013). Effects of high pressure treatment and temperature on lipid oxidation and fatty acid composition of yak (*Poephagus grunniens*) body fat. *Meat Science, 94*(4), 489–494.

Wang, R., Ding, S., Hu, X., Liao, X., & Zhang, Y. (2016). Effects of high hydrostatic pressure on chlorophylls and chlorophyll–protein complexes in spinach. *European Food Research and Technology, 242*(9), 1533–1543.

Wolbang, C. M., Fitos, J. L., & Treeby, M. T. (2008). The effect of high pressure processing on nutritional value and quality attributes of *Cucumis melo* L. *Innovative Food Science & Emerging Technologies, 9*(2), 196–200.

Yagiz, Y., Kristinsson, H. G., Balaban, M. O., & Marshall, M. R. (2007). Effect of high pressure treatment on the quality of rainbow trout (*Oncorhynchus mykiss*) and mahi mahi (*Coryphaena hippurus*). *Journal of Food Science, 72*(9), C509–C515.

Yagiz, Y., Kristinsson, H. G., Balaban, M. O., Welt, B. A., Ralat, M., & Marshall, M. R. (2009). Effect of high pressure processing and cooking treatment on the quality of Atlantic salmon. *Food Chemistry, 116*(4), 828–835.

Yang, H., Khan, M. A., Yu, X., Zheng, H., Han, M., Xu, X., & Zhou, G. (2016). Changes in protein structures to improve the rheology and texture of reduced-fat sausages using high pressure processing. *Meat Science, 121*, 79–87.

Ye, A., Anema, S. G., & Singh, H. (2004). High-pressure-induced interactions between milk fat globule membrane proteins and skim milk proteins in whole milk. *Journal of Dairy Science, 87*(12), 4013–4022.

Yi, J., Kebede, B. T., Hai Dang, D. N., Buvé, C., Grauwet, T., Van Loey, A., … Hendrickx, M. (2017). Quality change during high pressure processing and thermal processing of cloudy apple juice. *LWT, 75*, 85–92.

Yi, J., Feng, H., Bi, J., Zhou, L., Zhou, M., Cao, J., & Li, J. (2016). High hydrostatic pressure induced physiological changes and physical damages in asparagus spears. *Postharvest Biology and Technology, 118*, 1–10.

Zamri, A. I., Ledward, D. A., & Frazier, R. A. (2006). Effect of combined heat and high-pressure treatments on the texture of chicken breast muscle (*Pectoralis fundus*). *Journal of Agricultural and Food Chemistry, 54*(8), 2992–2996.

Zhang, C., Trierweiler, B., Li, W., Butz, P., Xu, Y., Rüfer, C. E., … Zhao, X. (2011). Comparison of thermal, ultraviolet-c, and high pressure treatments on quality parameters of watermelon juice. *Food Chemistry, 126*(1), 254–260.

Zhang, F., Dong, P., Feng, L., Chen, F., Wu, J., Liao, X., & Hu, X. (2012). Textural changes of yellow peach in pouches processed by high hydrostatic pressure and thermal processing during storage. *Food and Bioprocess Technology, 5*(8), 3170–3180.

Zhao, G., Zhang, R., & Zhang, M. (2017). Effects of high hydrostatic pressure processing and subsequent storage on phenolic contents and antioxidant activity in fruit and vegetable products. *International Journal of Food Science & Technology, 52*(1), 3–12.

Zhou, C.-L., Liu, W., Zhao, J., Yuan, C., Song, Y., Chen, D., … Li, Q.-H. (2014). The effect of high hydrostatic pressure on the microbiological quality and physical–chemical characteristics of Pumpkin (*Cucurbita maxima* Duch.) during refrigerated storage. *Innovative Food Science & Emerging Technologies, 21*, 24–34.

Chapter 5
Impact of Pulsed Light on Food Constituents

Peng Peng, Paul Chen, Dongjie Chen, Min Addy, Yanling Cheng, Nan Zhou, Charles Schiappacasse, Yaning Zhang, Erik Anderson, Juer Liu, Yiwei Ma, and Roger Ruan

5.1 Introduction

As one of the emerging non-thermal food treatment technologies, pulsed light (PL) has been intensively studied since the twenty first century. PL technology uses intermittent light pulses to treat food products without leaving any residues. The current use of PL technology is for decontamination purposes. The decontamination effects of PL treatment rely on primarily light with the different wavelength and the pulsed energy. Therefore, this chapter starts by explaining the disinfection mechanism of PL technology, where analysis of the UV disinfection mechanism would be helpful. Unlike electron beam, X-rays, and gamma rays, UV light is non-ionizing irradiation and does not break molecular bonds. The UV light can be emitted either as a continuous wave (continuous light) or in short duration pulses of 1–20 per second as in the case of pulsed light. Continuous UV light may be categorized into three types according to the emission spectrum: (1) short-wave UV (UV-C) with wavelengths from 200 to 280 nm, (2) medium-wave UV (UV-B) with wavelengths from 280 to

P. Peng · P. Chen · D. Chen · M. Addy · Y. Cheng · N. Zhou · C. Schiappacasse · E. Anderson
J. Liu · Y. Ma
Center for Biorefining, and Department of Bioproducts and Biosystems Engineering, University of Minnesota Twin Cities, St. Paul, MN, USA

Y. Zhang
Center for Biorefining, and Department of Bioproducts and Biosystems Engineering, University of Minnesota Twin Cities, St. Paul, MN, USA

The Harbin Institute of Technology, Haerbin Shi, Heilongjiang Sheng, China

R. Ruan (✉)
Center for Biorefining, and Department of Bioproducts and Biosystems Engineering, University of Minnesota Twin Cities, St. Paul, MN, USA

Nanchang University, Nanchang, China
e-mail: Ruanx001@umn.edu

S. Roohinejad et al. (eds.), *Effect of Emerging Processing Methods on the Food Quality*,
https://doi.org/10.1007/978-3-030-18191-8_5

320 nm, and (3) long-wave UV (UV-A) with wavelengths from 320 to 400 nm. UV-C has been used for disinfection of air, surface, water, and liquid foods (Bintsis, Litopoulou-Tzanetaki, & Robinson, 2000). The mechanisms responsible for microbial inactivation by UV light are believed to be due to the photochemical, photothermal and photophysical effects on microbes exposed to UV light (Oms-Oliu, Martín-Belloso, & Soliva-Fortuny, 2010).

Continuous UV light has a shallow penetration depth and is suitable for disinfection of transparent liquid such as water. For liquid foods such as juice and milk, the penetration depth is less, depending on the type of liquid. UV light has been used to treat fresh fruits and vegetables to reduce the initial count of microorganisms on the surface of the product and to induce host resistance to the microorganisms through a mechanism termed "hormesis" (Guerrero-Beltrán & Barbosa-Cánovas, 2004). UV-C has also been used to treat meat, poultry, fish, chocolate, and eggs to reduce microbial loads and extend shelf life without negatively affecting the quality (Djenane, Sánchez-Escalante, Beltrán, & Roncalés, 2001; Huang & Toledo, 1982; Kuo, Carey, & Ricke, 1997; Lee, Kermasha, & Baker, 1989; Wallner-Pendleton, Sumner, Froning, & Stetson, 1994).

Despite the demonstration of its apparent efficacy under laboratory conditions, UV disinfection technology in the food processing industries has seen few practical applications. Gardner and Shama (2000) attributed the phenomenon to two major reasons. First, there may be wrongly perceived undesirable changes to both the appearance and nutritional value of UV-treated foods, which may have deterred some potential users, despite the fact that most deleterious effects previously reported tending to be significant only at high UV fluencies. Second, the range of commercially available equipment for disinfecting solid foods is limited.

PL shares the same germicidal mechanisms as UV light. However, to a certain extent, PL overcomes the two major disadvantages of UV light over ionizing irradiation. First, PL has much greater penetration depth than UV light. Second, PL releases a peak power distribution during a pulse as high as 35 MW compared with 100–1,000 W for continuous UV light (Oms-Oliu, Martín-Belloso, & Soliva-Fortuny, 2010). The high penetration and energy level makes PL an excellent alternative to continuous UV light for effective inactivation of pathogens in foods (Birmpa, Vantarakis, Paparrodopoulos, Whyte, & Lyng, 2014). Cheigh, Park, Chung, Shin, and Park (2012) compared the effect of UV-C and PL on the viability and cell damage of *Listeria monocytogenes* and *Escherichia coli* O157:H7. UV-C irradiation for 1200 s resulted in a 4-log reduction of *L. monocytogenes* and a 5-log reduction of *E. coli* O157:H7 while PL treatments at energy densities of 376 and 455 W/m^2 for 60–180 s caused an approximately 7-log reduction in *L. monocytogenes* and *E. coli* O157:H7. The transmission electron microscopy study indicates that bacterial cell structures were destroyed by intense pulsed light (IPL) treatment but not by UV-C treatment. This was also observed in a study involving *L. monocytogenes* inoculated on solid medium and seafood (Cheigh, Hwang, & Chung, 2013). The mechanism of IPL on killing microbes was concluded as follows, (1) Photochemical damage: the UV light portion of the PL damages the DNA of bacteria by forming thymine dimers. Upon dimer formation, bacterial DNA cannot be unzipped for replication, and thus bacteria cannot reproduce. (2) Photothermal damage: localized heating of bacteria can be generated by PL treatment based on different thermal properties within the food-bacteria system. The thermal stress caused by this irregular heating would damage the cell via cell wall rupture and membrane decomposition (Keener & Krishnamurthy, 2014). (3) Photo-physical damage: PL is also expected to induce some physical disruption on microbial cellular structures due to the intermittent, high-intensity pulses.

Furthermore, PL not only kills vegetative bacteria but also inactivates spores (Nicorescu et al., 2013), which is usually difficult even with thermal processes. Fine and Gervais (2004) reported that during PL processing, oxidative reactions were limited because of the short pulse duration, typically 300 ns to 1 ms, compared with >10 min for continuous UV light. Finally, PL treatment does not leave residual compounds or use external chemical disinfectants and preservatives. Application of IPL in the food industry has been approved by the FDA under the code 21CFR179.41 (FDA, 1996).

5.2 Influences of Pulsed Light on Foods' Qualities

The primary purpose of using PL to treat food products is to inactivate microbes, and the most commonly researched food process involving PL treatment is shelf life extension. In this section, the effects of PL treatment in microbial disinfection and shelf life extension of fresh produces, meat products, and dairy products will be discussed. Furthermore, due to the photochemical, photothermal, and photophysical treatments by the PL, the changes in other foods' qualities may occur, which will also be reviewed in this section. In particular, the food qualities that will be discussed in this section are flavor, texture, color, nutrient contents, and sensory attributes. These qualities are crucial to the commercial applications of this technology and are critical factors in its scale-up evaluations.

5.2.1 Fruits and Vegetables

Several of the most important consumer expectations for the quality of fruits and vegetables are the appearance, texture, nutritional value, and shelf life stability. For fresh produce, especially the minimally processed foods, the greatest limitations for the extended shelf life is yeast proliferation (Gorny, 2003). Therefore, relevant studies used the yeast population as one critical method to determine the shelf life enhancement of the PL treatment. Studies prior to the year of 2010 demonstrated a maximum storage time of 9 days for PL-treated fresh produce, with 3 days being the time where obvious textual changes were observed (Gómez-López, Devlieghere, Bonduelle, & Debevere, 2005). One representative study was conducted by Oms-Oliu, Aguiló-Aguayo, Martín-Belloso, and Soliva-Fortuny (2010) who used fresh-cut mushrooms as an example to demonstrate the influence of PL treatment on the shelf life and antioxidant levels to fresh produces. The yeast populations on the treated mushrooms were analyzed compared with the untreated samples. Subjected to PL treatments between 4.8 and 28 J/cm^2, the yeast proliferation rate was reduced by over 1 log during storage. This study concluded that the PL treatment of fresh cut mushrooms led to an increase in the shelf life of the mushroom for 2 to 3 days. While this was a significant improvement compared with previous studies, this research also showed a decrease in nutrients including vitamins and antioxidants, which agreed with other later studies.

Table 5.1 summarizes the representative studies that reported the quality changes to PL-treated fruits and vegetables after the year of 2010. For the changes reported in the studies listed in Table 5.1, it was obvious that the nutritional quality changes caused by the PL treatment varied between different fruits and vegetables. Overall, the changes in the nutritional properties under PL treatment are not significant. The slight reduction in the content of phenolic compounds could be due to the break of the membrane cell walls; leading to higher activities of polyphenol oxidase resulted from the tissue damage and the loss of the functional cell compartmentalization (Gómez et al., 2012). From Table 5.1, it could be also observed that the effect of PL treatment on the textural quality changes was more obvious for fresh produces with smooth surfaces, such as apples and tomatoes. The increases in brownness dehydration of apples subjected to the PL treatment were repeatedly reported in the related studies (Gómez et al., 2012; Ignat et al., 2014; Moreira, Álvarez, Martín-Belloso, & Soliva-Fortuny, 2017). On the other hand, the fruits with rough surface experienced fewer changes in color and firmness when subjected to PL treatments. The literature reported the protective effects of the rough surfaces on microbes during the PL treatment (Maftei, Ramos-Villarroel, Nicolau, Martín-Belloso, & Soliva-Fortuny, 2014; Nicorescu et al., 2013). Therefore, it was reasonable to propose that the hindrance effects caused by the rough surface of the fruits and vegetables could provide protection to the textural properties as well. The shielding effect from the rough surfaces was a double-sided sword for the PL treatment as it could limit the decontamination performance while preserving the textural properties. Additionally, it could also be determined from these studies that the textural changes were more likely due to the dehydration effect caused by the visible and inferred portion of the PL treatment. For studies that implement solely the pulsed UV treatments, the fruits experienced no to little textual changes (Luksiene et al. 2013).

Although PL is considered as a non-thermal food treatment technology, it still causes temperature rise due to the energy dissipation in the process. Although in most of the cases, the temperature during PL treatment does not exceed 60 °C, certain destruction of nutritional (e.g. vitamins) and sensory attributes would still occur due to the thermal effects (Oms-Oliu, Aguiló-Aguayo, et al., 2010). Therefore, researchers have developed an innovative modification to the technology, called water-assisted PL or wet PL, to reduce the thermal damage of PL and preserve the foods' qualities, especially for fruits and vegetables that are sensitive to temperature rise (Bhavya & Umesh Hebbar, 2017).

This developed a novel modification that has been used in the laboratory scale to treat various fruit products such as blueberries and raspberries (Huang & Chen, 2014,

Table 5.1 The effect of pulsed light treatment on the quality of fruits and vegetables

Food matrix	Treatment conditions: Energy	Treatment conditions: Pulse info and treatment time	Best reported decontamination results	Nutritional properties	Textual properties	Sensorial properties	Reference
Apple	2.4–221.1 J/cm^2	360 μs pulse width for 2–100 s	2.25 logs for *E. coli*; 2.7 logs for *L. innocua*	Degraded walls and broken plasmalemma and tonoplast were observed	Apples turned darker and less green with the increased PL doses	N/A	Gómez, Salvatori, García-Loredo, and Alzamora (2012)
	1.75–15.75 J/cm^2	112 μs pulse width for up to 9 pulses	3 logs for *L. brevis* and *L. monocytogenes*	N/A	Dehydration and browning were observed	Negative changes in the flavor profile	Ignat, Manzocco, Maifreni, Bartolomeoli, and Nicoli (2014)
Strawberries	0–0.195 J/cm^2 (UV dose)	112 μs pulse width for 2–30 s	2.2 logs for mesophilic bacteria; 1.5 logs for *B. cereus*; 1.1 logs for *L. monocytogenes*	No significant changes in total phenolic, anthocyanin ascorbic acid content, and antioxidant capacity	No impact on the color and firmness	N/A	Luksiene, Buchovec, and Viskelis (2013)
Raspberries	0–28.2 J/cm^2	3 pulse/s for 5, 15, and 30 s	4.5 logs for most Salmonellas; 3.9 logs for *E. coli*	No changes in the total phenolic contents	The softer texture was observed during the storage	N/A	Xu and Wu (2016)
Tomato fruit	2.56–5.36 J/cm^2	150 μs pulse width for 1–2 pulses	2.3 logs for *S. cerevisiae*	No change in nutritional quality as well as vitamin C content; Slight increase in carotenoid concentrations	Partial dehydration, weightloss, and wrinkleness were observed	N/A	Aguiló-Aguayo, Charles, Renard, Page, and Carlin (2013)

2015; Huang, Sido, Huang, & Chen, 2015). In these studies, there have been no changes in the fruits' appearance and the surface heating has been greatly reduced. Therefore, wet PL treatment, or water-assisted PL treatment, could be one of the promising PL processes used in the fruit and vegetable industry in the future.

5.2.2 Meat Products

While there were many studies related to the treatment of fruits and vegetables using PL technology, investigations on the use of PL for meat treatment were not as thorough. However, similar to the studies on fruits and vegetables, the primary application of PL treatment remained to be the surface decontamination and extension of shelf lives. Related studies targeted the treatment of two meat products, raw meat, and ready-to-eat meat. For the investigations on the fresh meat, one of the early studies conducted by Ozer and Demirci (2006) reported the inactivation of *E. coli* and *L. monocytogenes*. In this study, the reduction achieved for both microbes were slightly less than or equal to 1-log, which was not as sufficient as the ones achieved for fruits and vegetables (Gómez et al., 2012; Ignat et al., 2014). In a later study conducted by Hierro, Ganan, Barroso, and Fernández (2012), a comprehensive inactivation of various microbes on beef and tuna by PL, including *L. monocytogenes*, *E. coli*, *S. Typhimurium* and *V. parahaemolyticus* was reported. In this study, approximately 1-log reduction was achieved for the above-mentioned contaminants. Although improvements on the disinfection performances were shown in this study compared with Ozer and Demirci (2006), the authors stated that for fresh meat products, the results still required improvements to completely reduce cross-contamination within the production facilities. Overall, although in the field of meat treatment with the PL technology, the studies on the treatment of fish products were reported more intensively than other products, the studies on products such as beef, pork, and chicken were reported. For the PL treatments on beef, a similar log-reduction result was obtained as the treatment for tuna (Hierro et al., 2012). The results for the decontamination of beef were less effective than the treatment of pulsed UV light on chicken meat (Paškevičiūtė & Lukšienė, 2009). Also, the difference in UV doses might contribute to the changes in the disinfection results and the different surface textural properties of these two types of meat could be the primary reason why the results varied. Furthermore, the decontamination of *S. aureus* on fermented salami, together with the reported *L. monocytogenes*, *E. coli*, *S.* Typhimurium and *V. parahaemolyticus* was recently reported (Rajkovic, Tomasevic, De Meulenaer, & Devlieghere, 2017). The authors found that the inactivation of these contaminants led to over 1-log additional reduction for tuna (Hierro et al., 2012), with higher energy doses (15 J/cm^2 for salami and 11.2 J/cm^2 for tuna).

For the sensory and nutritional qualities, the changes caused by PL treatment on meats were reported. For instance, for the fermented salami, PL treatment caused a notable increase in protein oxidation and carbonyl content (Rajkovic et al., 2017). It was not surprising to see this finding as the negative effects of PL treatment on meat products have been reported (Nicorescu, Nguyen, Chevalier, & Orange, 2014; Tomašević, 2015). Nicorescu et al. (2014) pointed out in their study that the 30 J/cm^2 of PL dose was the value that led to significant color and sensory quality changes. At this PL dose, malondialdehyde content in the roasted pork increased by up to 39.3%.

On the other hand, one study concluded that PL treatment caused less damage to the dry cured meat than the cooked or raw meat (Tomašević & Rajković, 2015). This result agreed with a previous study on the use of PL to decontaminate dry-cured meat products. For the cooked and ready-to-eat meat used in this study, almost no significant changes in the sensory qualities of the meat were observed after the products were treated by a PL dose of 11.9 J/cm^2 (Ganan, Hierro, Hospital, Barroso, & Fernández, 2013). The study reported that the small changes in sensory qualities disappeared during storage and did not affect the overall evaluation scores by the consumer panelists for these ready-to-eat meats.

To conclude, the studies on using PL to treat meat products were scattered. Although the effect of PL treatment on different meat products was reported, they were not intensively studied as the fruits and vegetables. Furthermore, the extension of the shelf lives of the meat products was also less than those of fresh produces. To examine this, more studies should be conducted with regards to meat surface and microbial characterizations. PL treatment led to negative impacts on the quality and sensory properties of raw meat but negligible effect on those of cooked and dry cured meat, and the difference of the changes became less significant after storage.

5.2.3 Dairy Products

Research on applying PL technology to dairy products began in the late 1990s and early 2000s when the decontamination effects of PL treatment to *Pseudomonas*, *E. coli*, and *Salmonella* were tested on cottage cheese and milk (Dunn, 1995; Smith, Lagunas-Solar, & Cullor, 2002). Overall, studies on PL treatment of dairy products were not performed as much as fruits and vegetables. During the past ten years, the studies related to the PL treatment of dairy products remained to focus on cheese and milk. However, other than the decontamination effects, recent studies started to investigate the nutritional quality changes of the milk and cheese.

For milk products, the use of PL treatment was effective in terms of shelf life extension (Rysstad & Kolstad, 2006). PL treatment demonstrated promising disinfection performances for various bacteria in different kinds of milk such as concentrated milk, regular milk, and goat milk (Kasahara, Carrasco, & Aguilar, 2015; Miller, Sauer, & Moraru, 2012). On the other hand, the inactivation of spores in milk by PL treatment was sparsely reported. On the nutritional quality point of view, the effects of the PL treatment on the protein and lipid contents in milk were also investigated. Elmnasser et al. (2008) stated in their publication that PL and UV-PL treatment caused no significant change to the protein and lipid contents. For sensory properties, Kasahara et al. (2015) found no differences between the PL-treated and the original goat milk samples. However, negative flavor and aroma changes were only observed repeatedly for the ones treated by continuous or pulsed UV treatments (Orlowska et al., 2012; van Aardt et al., 2005). Therefore, it was reasonable to propose that the photochemical effects from the UV irradiation were the main causes of the changes in the proteinogenic amino acids, and in lipids, which further led to the deterioration in the sensory properties.

For cheese products, the disinfection performances of the *P. roqueforti* and *L. monocytogenes* were evaluated in addition to the previously mentioned *E. coli* strain (Can, Demirci, Puri, & Gourama, 2014). In this study, the greatest log reductions observed for *P. roqueforti* and *L. monocytogenes* were 1.24 and 2.9 logs, respectively. Although the results were not as efficient as *E. coli*, PL treatment did not cause significant changes to the color and chemical properties of the cheese. Note that in this study, a slight decrease in the elastic modulus was observed in the PL-treated cheese. Yet the changes were not sufficient enough to affect the overall mechanical properties of the cheese. A later study carried out by Fernández, Ganan, Guerra, and Hierro (2014) pointed out that the PL treatment with less than 4.8 J/cm^2 did not change the qualities while treatments greater than 8.4 J/cm^2 led to a moderate or significant increase in the protein oxidation. Moreover, the sensory evaluation of the PL-treated cheese was conducted in another study by Fernández, Hospital, Arias, and Hierro (2016). The results from this study confirmed the upper treatment limit of around 4–5 J/cm^2 to avoid the significant quality changes. However, the changes in the sulfur volatiles and the related sensory changes gradually disappeared during cold storage. Therefore, PL treatment could be a potential useful decontamination tool in the cheese industry for the products that would be subjected to cold storage after the treatment.

In addition to cheese and milk, the PL technology was recently applied to the whey protein extract (Siddique, Maresca, Pataro, & Ferrari, 2016). Application of PL treatment was reported to improve the foaming and solubility of the whey protein extracts via dissociation and partial unfolding of the protein isolate. This study was important to the PL technology since it proved the feasibility of using PL technology to improve the properties of functional dairy products and led to an avenue of future applications of PL technology in functional foods.

5.3 Challenges

5.3.1 Challenges as a Surface Treatment Technology

Like other surface treatment technologies, one of the major challenges faced by PL technology is the lack of deep penetration. Although the penetration depth of the PL treatment is greater compared with continuous UV due to the enhancement in the pulse energy, the penetration is still lower than the conventional thermal treatment or other types of irradiations. The shielding effects caused by the complex surfaces limits the disinfection power of PL treatment on many food matrices. Furthermore, PL treatment might not be sufficient enough to inactivate bacteria that are hidden inside the cavities and irregular surfaces of different foods (Elmnasser et al., 2007). Within the same food particles, the different surface textural properties might lead to various inactivation susceptibilities. Because PL is a surface treatment technology, it would also be less effective when used to kill the microorganisms absorbed within the complex structure of certain foods, such as fat in meat products, and fail to extend the shelf life of these foods.

For foods with larger particle sizes, the incomplete exposure of the entire food to PL is another challenge. Water-assisted PL treatment could potentially reduce this problem by submerging the food into the water to enhance light exposure (Huang, Sido, Huang, & Chen, 2015). However, the water-assisted apparatus still requires further improvements for it to become more efficient and commercially feasible. Lastly, for PL treatments, the fluidized bed types of conveying systems can lead to more complete and uniform treatment of food products. Therefore, the competency of the other food transport systems, such as the conveyor belt, is another challenge for the commercial application of PL technology from the penetration and exposure problems.

5.3.2 *Effect on the Photochemical and Photothermal Compounds*

The quality changes described in Sect. 5.2 are mostly due to the photochemical and photothermal effects of PL treatment. The photochemical effects lead to chemical reactions that could cause oxidation and reduce the nutrients of the food ingredients. Therefore, the industrial applications of PL technology in food products might need to be conducted in nitrogen (or other inert gases) environment. Photothermal effects of the PL treatment can also reduce the nutrient content and sensory qualities of the foods (Bhavya & Umesh Hebbar, 2017). Although PL technology is considered as a non-thermal technology, the temperature rises in the food being treated must be controlled by adjusting the process parameters (exposure time, pulse energy, frequency, particle size, distance to the light, etc.). Due to this complexity, the commercial application of PL technology is challenging to certain foods, especially solid foods with irregular shape, chemical, and physical properties. Not only the non-uniformity of the food will cause uncertainty in microbial inactivation, but the quality and sensory changes in certain parts of the food might also be lowered to unacceptable levels.

5.4 Future Opportunities

For decontamination and shelf life extension purposes, one of the major opportunities for the PL treatment will be for the powdered foods. Powdered foods are widely used as the ingredients in manufacturing processed foods or consumed directly by humans and animals for their energy and nutrient contents. The popularity of powdered foods is rising due to the convenience and versatility of their use in food preparation. Inappropriate and insufficient decontamination has led to numerous outbreaks of foodborne diseases in recent years due to the existence of pathogenic microbes in dry milk powder, infant formula, spices, bread crust, etc., or through the cross-contamination when inappropriately pasteurized food ingredients such as spices were added into meat, pasta and pizza products. In recent years, there have been several major disease outbreaks in powdered foods around the world (Brandt, Serrano Oria, Kallon, & Bazzano, 2017; Jourdan-da Silva et al., 2018; Mullen, 2017). Due to their low water activity (a_w), powdered foods pose a unique challenge for disinfection. Many spores forming-microbes, including *Clostridium spp.* and *Bacillus spp.* adapt to the low a_w hostile environment by becoming physiologically dormant and metabolically quiescent, rendering them less susceptible to bactericidal interventions.

Different physical and chemical processes have been used to decontaminate powdered foods. Thermal treatments are commonly used, as they are easier to implement than others. Dry/powdered foods are commonly pasteurized thermally in the liquid state, for example, 72 °C for 15 s is used for milk disinfection before evaporation for making milk powder products. Thermal treatments are commonly used, as they are easier to implement than others. To prevent post-packaging recontamination, thermal treatment is often applied to pre-packaged foods. As a result, the treatment has to be so severe that the outer portion of the food is overheated while the center is still cold. For example, dry powdered foods are usually packed in 30–40 lb cartons. A pallet of 20 cartons is heat-treated in a chamber at 90 °C (195 °F). After a 20–30 min treatment, the center of the stack cannot reach the inactivation temperature. This will adversely affect the quality and safety of the treated products. Furthermore, microbes exhibit greater thermal resistance in dry environments than in more humid conditions. The high-temperature steam treatment was found to decrease volatile oil content, degrade color, and increase the moisture content of the dried spices, which leads to a decreased shelf life (Lilie, Hein, Wilhelm, & Mueller, 2007).

Inactivation of dry foods by ionizing irradiation using energy such as electron beam, X-rays, and gamma rays has been reported (Niemira, 2014). The electron beam has a much shorter penetration depth than X-rays and gamma rays and therefore the latter two techniques are more suitable for processing of pallet/crate loads of foods. Irradiation of dry herbs and spices is very common in the US, with 78% of 103,000 tons herbs and spices that are irradiated annually. Many defects of these processes have been identified, including the lack of efficiency, difficulty to perform, and production of undesirable side products (the formation of potential mutagens and carcinogens in chemical methods) (Steenland, Whelan, Deddens, Stayner, & Ward, 2003).

The PL technology is also suitable for foods with small particle sizes. As mentioned in Sect. 5.3.1, the two major challenges for PL technology are the low penetration depth and uniform treatment to large solid foods. Powdered foods circumvent these challenges due to their small particle size and greatly reduce the shielding effect caused by large solid foods. By using a fluidized conveying system, they could be continuously and uniformly treated by PL technology, and most importantly, without introducing moisture or heat. Similar to powdered foods, fluid foods can reduce the shielding effects and enhance the penetration of the PL treatment. Furthermore, for most functional fluid foods, the physical and chemical properties are uniform at the final stages of the production line. Therefore, the dosage of PL treatment could be adjusted according to the compositions to reach certain desired function changes.

5.5 Conclusions

In this chapter, the effects of PL treatment on the microbial, physicochemical, nutritional, and sensorial properties of fruits and vegetables, meat products, and dairy products were introduced. Among the three products, the applications of PL technology on fruits and vegetables have been extensively investigated, whereas the study of the meat and dairy products were relatively less documented. PL treatment was successful in extending the shelf life of all three kinds of food products. In general, the microbial inactivation and shelf life extension performances were better demonstrated in the products with smooth surfaces. For nutritional and sensory qualities, the most common effects caused by PL on fresh produces were the dehydration and color changes, while no significant sensory changes occurred. For fresh meat products, the increase in protein oxidation and carbonyl content were the main factors that led to the negative impacts of PL treatment on sensory and nutrient qualities. However, the impacts were significantly less for cooked meat and had a tendency to disappear during storage. Similarly, for dairy products such as milk, the low dose PL treatments (below 4 to 5 J/cm^2) led to no significant changes in protein and lipid content and sensory qualities. Furthermore, the changes would also disappear during cold storage. The two challenges identified for PL treatment were the relatively short penetration depth and the potentially negative impacts caused by PL's photochemical and photothermal effects. These challenges can be overcome by controlling the process parameters such as exposure time and pulse energy. Based on the challenges, PL technology has the potential to play a key role in treating powdered foods and functional fluid foods in the future. These two products circumvent the challenges of PL. The small particle size of the powdered food and the greater transparency of the functional drinks allow full exposure of the products once they are fluidized. The fluidization of these products is also easier to achieve than large solid foods. Lastly, the uniformity of these foods avoids the partial overheat during PL treatment, which is the main limitation for foods with complex surfaces and large particle sizes.

References

Aguiló-Aguayo, I., Charles, F., Renard, C. M. G. C., Page, D., & Carlin, F. (2013). Pulsed light effects on surface decontamination, physical qualities and nutritional composition of tomato fruit. *Postharvest Biology and Technology, 86*, 29–36.

Bhavya, M. L., & Umesh Hebbar, H. (2017). Pulsed light processing of foods for microbial safety. *Food Quality and Safety, 1*(3), 187–202.

Bintsis, T., Litopoulou-Tzanetaki, E., & Robinson, R. K. (2000). Existing and potential applications of ultraviolet light in the food industry – a critical review. *Journal of the Science of Food and Agriculture, 80*(6), 637–645.

Birmpa, A., Vantarakis, A., Paparrodopoulos, S., Whyte, P., & Lyng, J. (2014). Efficacy of three light technologies for reducing microbial populations. *BioMed Research International.* https://doi.org/10.1155/2014/673939

Brandt, A., Serrano Oria, Ó., Kallon, M., & Bazzano, A. N. (2017). Infant feeding policy and programming during the 2014–2015 ebola virus disease outbreak in Sierra Leone. *Global Health: Science and Practice, 5*(3), 507–515.

Can, F. O., Demirci, A., Puri, V. M., & Gourama, H. (2014). Decontamination of hard cheeses by pulsed UV light. *Journal of Food Protection, 77*(10), 1723–1731.

Cheigh, C.-I., Hwang, H.-J., & Chung, M.-S. (2013). Intense pulsed light (IPL) and UV-C treatments for inactivating *Listeria monocytogenes* on solid medium and seafoods. *Food Research International, 54*(1), 745–752.

Cheigh, C.-I., Park, M.-H., Chung, M.-S., Shin, J.-K., & Park, Y.-S. (2012). Comparison of intense pulsed light- and ultraviolet (UVC)-induced cell damage in *Listeria monocytogenes* and *Escherichia coli* O157:H7. *Food Control, 25*(2), 654–659.

Djenane, D., Sánchez-Escalante, A., Beltrán, J. A., & Roncalés, P. (2001). Extension of the retail display life of fresh beef packaged in modified atmosphere by varying lighting conditions. *Journal of Food Science, 66*(1), 181–186.

Dunn, J. (1995). Pulsed-light treatment of food and packaging. *Food Technology, 49*, 95–98.

Elmnasser, N., Dalgalarrondo, M., Orange, N., Bakhrouf, A., Haertlé, T., Federighi, M., & Chobert, J.-M. (2008). Effect of pulsed-light treatment on milk proteins and lipids. *Journal of Agricultural and Food Chemistry, 56*(6), 1984–1991.

Elmnasser, N., Guillou, S., Leroi, F., Orange, N., Bakhrouf, A., & Federighi, M. (2007). Pulsed-light system as a novel food decontamination technology: A review. *Canadian Journal of Microbiology, 53*(7), 813–821.

FDA. (1996). *Code of federal regulations. 21CFR179.41*. Retrieved from https://www.ecfr.gov/cgi-bin/text-idx?SID=980dfb81a14d4d705047084321e162b8&mc=true&tpl=/ecfrbrowse/Title21/21cfr179_main_02.tpl

Fernández, M., Ganan, M., Guerra, C., & Hierro, E. (2014). Protein oxidation in processed cheese slices treated with pulsed light technology. *Food Chemistry, 159*, 388–390.

Fernández, M., Hospital, X. F., Arias, K., & Hierro, E. (2016). Application of pulsed light to sliced cheese: Effect on *Listeria* inactivation, sensory quality and volatile profile. *Food and Bioprocess Technology, 9*(8), 1335–1344.

Fine, F., & Gervais, P. (2004). Efficiency of pulsed UV light for microbial decontamination of food powders. *Journal of Food Protection, 67*(4), 787–792.

Ganan, M., Hierro, E., Hospital, X. F., Barroso, E., & Fernández, M. (2013). Use of pulsed light to increase the safety of ready-to-eat cured meat products. *Food Control, 32*(2), 512–517.

Gardner, D. W., & Shama, G. (2000). Modeling UV-induced inactivation of microorganisms on surfaces. *Journal of Food Protection, 63*(1), 63–70.

Gómez, P. L., Salvatori, D. M., García-Loredo, A., & Alzamora, S. M. (2012). Pulsed light treatment of cut apple: Dose effect on color, structure, and microbiological stability. *Food and Bioprocess Technology, 5*(6), 2311–2322.

Gómez-López, V. M., Devlieghere, F., Bonduelle, V., & Debevere, J. (2005). Intense light pulses decontamination of minimally processed vegetables and their shelf-life. *International Journal of Food Microbiology, 103*(1), 79–89.

Gorny, J. R. (2003). A summary of CA and MA requirements and recommendations for fresh-cut (minimally processed) fruits and vegetables. *Acta Horticulturae, 600*, 609–614.

Guerrero-Beltrán, J. A., & Barbosa-Cánovas, G. V. (2004). Review: Advantages and limitations on processing foods by UV light. *Food Science and Technology International, 10*(3), 137–147.

Hierro, E., Ganan, M., Barroso, E., & Fernández, M. (2012). Pulsed light treatment for the inactivation of selected pathogens and the shelf-life extension of beef and tuna carpaccio. *International Journal of Food Microbiology, 158*(1), 42–48.

Huang, Y., & Chen, H. (2014). A novel water-assisted pulsed light processing for decontamination of blueberries. *Food Microbiology, 40*, 1–8.

Huang, Y., & Chen, H. (2015). Inactivation of *Escherichia coli* O157:H7, *Salmonella* and human norovirus surrogate on artificially contaminated strawberries and raspberries by water-assisted pulsed light treatment. *Food Research International, 72*, 1–7.

Huang, Y., Sido, R., Huang, R., & Chen, H. (2015). Application of water-assisted pulsed light treatment to decontaminate raspberries and blueberries from *Salmonella*. *International Journal of Food Microbiology, 208*, 43–50.

Huang, Y.-W., & Toledo, R. (1982). Effect of high doses of high and low intensity UV irradiation on surface microbiological counts and storage-life of fish. *Journal of Food Science, 47*(5), 1667–1669.

Ignat, A., Manzocco, L., Maifreni, M., Bartolomeoli, I., & Nicoli, M. C. (2014). Surface decontamination of fresh-cut apple by pulsed light: Effects on structure, colour and sensory properties. *Postharvest Biology and Technology, 91*, 122–127.

Jourdan-da Silva, N., Fabre, L., Robinson, E., Fournet, N., Nisavanh, A., Bruyand, M., … Le Hello, S. (2018). Ongoing nationwide outbreak of *Salmonella* Agona associated with internationally distributed infant milk products, France, December 2017. *Euro Surveillance: Bulletin Europeen Sur Les Maladies Transmissibles = European Communicable Disease Bulletin, 23*(2). https://doi.org/10.2807/1560-7917.ES.2018.23.2.17-00852

Kasahara, I., Carrasco, V., & Aguilar, L. (2015). Inactivation of *Escherichia coli* in goat milk using pulsed ultraviolet light. *Journal of Food Engineering, 152*, 43–49.

Keener, L., & Krishnamurthy, K. (2014). Shedding light on food safety: Applications of pulsed light processing. *Food Safety Magazine, 20*, 28–33.

Kuo, F.-L., Carey, J. B., & Ricke, S. C. (1997). UV irradiation of shell eggs: Effect on populations of aerobes, molds, and inoculated *Salmonella* Typhimurium. *Journal of Food Protection, 60*(6), 639–643.

Lee, B. H., Kermasha, S., & Baker, B. E. (1989). Thermal, ultrasonic and ultraviolet inactivation of *Salmonella* in thin films of aqueous media and chocolate. *Food Microbiology, 6*(3), 143–152.

Lilie, M., Hein, S., Wilhelm, P., & Mueller, U. (2007). Decontamination of spices by combining mechanical and thermal effects – an alternative approach for quality retention. *International Journal of Food Science & Technology, 42*(2), 190–193.

Luksiene, Z., Buchovec, I., & Viskelis, P. (2013). Impact of high-power pulsed light on microbial contamination, health promoting components and shelf life of strawberries. *Food Technology and Biotechnology, 51*, 284.

Maftei, N. A., Ramos-Villarroel, A. Y., Nicolau, A. I., Martín-Belloso, O., & Soliva-Fortuny, R. (2014). Influence of processing parameters on the pulsed-light inactivation of *Penicillium expansum* in apple juice. *Food Control, 41*, 27–31.

Miller, B. M., Sauer, A., & Moraru, C. I. (2012). Inactivation of *Escherichia coli* in milk and concentrated milk using pulsed-light treatment. *Journal of Dairy Science, 95*(10), 5597–5603.

Moreira, M. R., Álvarez, M. V., Martín-Belloso, O., & Soliva-Fortuny, R. (2017). Effects of pulsed light treatments and pectin edible coatings on the quality of fresh-cut apples: A hurdle technology approach. *Journal of the Science of Food and Agriculture, 97*(1), 261–268.

Mullen, J. (2017). Baby formula recalled worldwide over salmonella scare. *CNN Money*.

Nicorescu, I., Nguyen, B., Chevalier, S., & Orange, N. (2014). Effects of pulsed light on the organoleptic properties and shelf-life extension of pork and salmon. *Food Control, 44*, 138–145.

Nicorescu, I., Nguyen, B., Moreau-Ferret, M., Agoulon, A., Chevalier, S., & Orange, N. (2013). Pulsed light inactivation of *Bacillus subtilis* vegetative cells in suspensions and spices. *Food Control, 31*(1), 151–157.

Niemira, B. A. (2014). Irradiation, microwave, and alternative energy-based treatments for low-water activity foods. In J. B. Gurtler, M. P. Doyle, & J. L. Kornacki (Eds.), *The Microbiological safety of low water activity foods and spices* (pp. 389–401). New York, NY: Springer.
Oms-Oliu, G., Aguiló-Aguayo, I., Martín-Belloso, O., & Soliva-Fortuny, R. (2010). Effects of pulsed light treatments on quality and antioxidant properties of fresh-cut mushrooms (*Agaricus bisporus*). *Postharvest Biology and Technology, 56*(3), 216–222.
Oms-Oliu, G., Martín-Belloso, O., & Soliva-Fortuny, R. (2010). Pulsed light treatments for food preservation. A review. *Food and Bioprocess Technology, 3*(1), 13–23.
Orlowska, M., Koutchma, T., Grapperhaus, M., Gallagher, J., Schaefer, R., & Defelice, C. (2012). Continuous and pulsed ultraviolet light for nonthermal treatment of liquid foods. Part 1: Effects on quality of fructose solution, apple juice, and milk. *Food and Bioprocess Technology, 6*(6), 1580–1592.
Ozer, N. P., & Demirci, A. (2006). Inactivation of *Escherichia coli* O157:H7 and *Listeria monocytogenes* inoculated on raw salmon fillets by pulsed UV-light treatment. *International Journal of Food Science & Technology, 41*(4), 354–360.
Paškevičiūtė, E., & Lukšienė, Ž. (2009). High-power pulsed light for decontamination of chicken breast surface. *Cheminė Technologija, 4*, 53.
Rajkovic, A., Tomasevic, I., De Meulenaer, B., & Devlieghere, F. (2017). The effect of pulsed UV light on *Escherichia coli* O157:H7, *Listeria monocytogenes*, *Salmonella* Typhimurium, *Staphylococcus aureus* and staphylococcal enterotoxin A on sliced fermented salami and its chemical quality. *Food Control, 73*, 829–837.
Rysstad, G., & Kolstad, J. (2006). Extended shelf life milk—advances in technology. *International Journal of Dairy Technology, 59*(2), 85–96.
Siddique, M. A. B., Maresca, P., Pataro, G., & Ferrari, G. (2016). Effect of pulsed light treatment on structural and functional properties of whey protein isolate. *Food Research International, 87*, 189–196.
Smith, W. L., Lagunas-Solar, M. C., & Cullor, J. S. (2002). Use of pulsed ultraviolet laser light for the cold pasteurization of bovine milk. *Journal of Food Protection, 65*(9), 1480–1482.
Steenland, K., Whelan, E., Deddens, J., Stayner, L., & Ward, E. (2003). Ethylene oxide and breast cancer incidence in a cohort study of 7576 women (United States). *Cancer Causes & Control, 14*(6), 531–539.
Tomašević, I. (2015). Intense light pulses upset the sensory quality of meat products. *Tehnologija Mesa, 56*, 1–7.
Tomašević, I., & Rajković, A. (2015). The sensory quality of meat, game, poultry, seafood and meat products as affected by intense light pulses: A systematic review. *Procedia Food Science, 5*, 285–288.
van Aardt, M., Duncan, S. E., Marcy, J. E., Long, T. E., O'Keefe, S. F., & Nielsen-Sims, S. R. (2005). Aroma analysis of light-exposed milk stored with and without natural and synthetic antioxidants. *Journal of Dairy Science, 88*(3), 881–890.
Wallner-Pendleton, E. A., Sumner, S. S., Froning, G. W., & Stetson, L. E. (1994). The use of ultraviolet radiation to reduce *Salmonella* and psychrotrophic bacterial contamination on poultry carcasses. *Poultry Science, 73*(8), 1327–1333.
Xu, W., & Wu, C. (2016). The impact of pulsed light on decontamination, quality, and bacterial attachment of fresh raspberries. *Food Microbiology, 57*, 135–143.

Chapter 6
Impact of Microwave Irradiation on Food Composition

Giancarlo Cravotto and Arianna Binello

6.1 Introduction

The use of high-frequency radio waves to heat and cook food dates back to the 1920s and the invention of vacuum tube radio transmitters. Microwaves (MW) fall between radio frequencies and infrared (IR) radiation frequencies in the electromagnetic spectrum and range between 0.3 GHz and 300 GHz with wavelengths ranging between 1 m and 1 cm. MW is an innocuous radiation type, when used with standard oven protection, and exert low vibrational energy that does not interact at atomic or molecular levels (Cravotto & Carnaroglio, 2017). In the early 1930s, the American magazine for radio experimenters published a popular editorial entitled "Cooking by ultrashort waves" (Gensback, 1933). This new and faster way of cooking was a great invention, which heated food without direct contact with hot surfaces, but only via irradiation across the path of the radio transmitter's power. In the late 1930s, two leading American companies; Westinghouse and Bell Laboratories, released several patent applications demonstrating the efficient cooking of foods by dielectric heating at ca. 60 MHz. The food industry rapidly recognized the huge potential that MW showed as a food processing technique (Puligundla et al., 2013), while physicists and engineers developed different versions of magnetrons that were used in military defense as radars. One of those researchers was Percy L. Spencer who, inspired by a serendipitous finding, attempted to heat all types of food with his device, leading to his company, Raytheon, filing the first patent to describe MW oven prototype a few years later (1945). Only at the end of the 1960s, domestic ovens became available for American families at an affordable price (less than 500 USD). Industrial applications of MW technology for food processing and drying have grown steadily since then and the frequencies of 2.45 GHz and 915 MHz became more common (Puligundla et al., 2013). Dielectric heating has several

G. Cravotto (✉) · A. Binello
Università di Torino, dipartimento di Scienza e Tecnologia del Farmaco, Torino, Italy
e-mail: giancarlo.cravotto@unito.it

S. Roohinejad et al. (eds.), *Effect of Emerging Processing Methods on the Food Quality*,
https://doi.org/10.1007/978-3-030-18191-8_6

advantages over conventional heating methods, especially with regards to energy efficiency. The peculiar type of volumetric heating that it provides dramatically reduces the temperature gradient between the outside and inside of food (Witkiewicz & Nastaj, 2010).

Although the development of robust theoretical and empirical models for process optimization is of paramount importance, the complex and variable composition of raw and pre-packaged foods makes this goal particularly challenging. Water, polar compounds and charged ions in food interact with MW electromagnetic fields intensely, causing the dipolar rotation and molecular friction that generate heat. MW heating and the dielectric properties of foods are influenced by temperature, moisture, salt content and radiation frequency (Buffler, 1993). Applications of MW for food processing typically include drying, tempering, sterilization, baking, and freezing, among others (Chizoba Ekezie, Sun, Han, & Cheng, 2017).

6.2 Advantages and Challenges of Using Microwave Radiation: Nutritional, Textural and Sensorial Properties

Several books, reviews, and articles have reported the use of MW technology in food processing, especially in drying, heating, cooking and sterilizing processes. A great deal of attention has very recently been paid to the existence of any changes in the quality of the food product after MW processing (Guo, Sun, Cheng, & Han, 2017).

6.2.1 Microwave Drying

Drying is a procedure that can be applied to food products for their preservation, storage, and transport. Physical changes and chemical modifications are often found to occur following this kind of treatment; alterations in color, flavor, textural properties, and nutritional value may affect the quality of the final product. For example, water removal leads to inevitable nutrient losses during the drying of vegetables (Oliveira, Brandão, & Silva, 2016). However, food safety must be preserved in all production stages and is the priority (Vadivambal & Jayas, 2007). The last decade has seen MW drying become a viable alternative to conventional hot air drying for a wide variety of food and agricultural products, as it allows drying time to be greatly reduced and quality degradation to be therefore limited.

This faster process, where high-frequency electromagnetic energy is converted into heat and intense liquid evaporation occurs, means that smaller floor spaces are required, making it possible to design more compact equipment (Li, Wang, & Kudra, 2011). Moreover, operational costs are reduced as heating occurs principally within the product and energy is not wasted in heating the walls of the oven. Avoiding heat loss means that fast start-up and shut-down are feasible using MW and that drying effects are more uniform (Sorour & El-Mesery, 2014), while the time saved over the total drying process can be reduced by up to 90% (Alibas, 2007). Maximum process efficiency occurs when the product to be dried presents low moisture content (less than 20%), as volumetric heating with MW then makes moisture removal quicker than it would be under hot air. However, at higher moisture content levels, MW drying can quickly give up its economic advantages (Mullin, 1994). Furthermore, the removal of moisture from food during MW drying can occur without case hardening occurring, while product quality is generally improved, even if foods can sometimes exhibit a more porous structure after treatment. There have been positive reports of MW drying leading to improvements in flexibility, color, flavor, nutritional and functional capacities, microbial stability, enzyme inactivation, and rehydration capacity (Dehnad, Jafari, & Afrasiabi, 2016; Nijhuis et al., 1998). While a number of results have shown that the quality of MW-treated products, such as grains, fruits, and vegetables, is either better or equal to that of conventionally dried products. Non-uniform heat distribution and higher equipment costs, however, may still present difficulties. A noteworthy improvement in β-carotene content was found during the dehydration of grated carrots when combinations of MW and vacuum, and drying under two-stage MW power (Arikan, Ayhan, Soysal, & Esturk, 2012). Vitamin C degradation when vegetables are dried is higher under hot-air drying (24–71%) than under MW-assisted processes, as the drying periods for the MW process are much shorter. As regards to flavor degradation, it is clear that any drying process will affect volatiles and flavor intensity and notes. Both convective and MW drying cause volatile losses especially in the early stages of drying when high temperatures are applied, at high drying temperatures and in foods with low moisture content (Oliveira et al., 2016). Vacuum MW drying was found to be an effective berry dehydration method as the absence of air accelerated evaporation and protected polyphenols from oxidative degradation (Li, Chen, Zhang, & Fu, 2017). The drying of spices and herbs under higher power MW energy has been found to give excellent product quality when very short treatment times were used (Kubra, Kumar, & Rao, 2016). Carefully designing the drying method, according to the nature, geometry and dielectric properties of the food matrix involved, is fundamental to guarantee minimum nutrient losses and either preserving or improving food quality. Although the use of MW-based drying systems in the food industry shows a great deal of potential, their limited penetration depth means that attention must be paid to scale up if this food-production potential is to be realized.

A novel approach to improving the functional properties of drying-process products can be found in combining MW, ultrasound, and conventional techniques. In fact, better quality, a lack of detrimental mechanical stresses in the products, and a maximum drying time reduction of up to 79% have all been achieved when this combined technique was used on raspberries (convective drying at 55 °C, MW drying at 100 W for 10 min, and ultrasound at 200 W) (Kowalski, Pawłowski, Szadzińska, Łechtańska, & Stasiak, 2016). Another novel technology in this field involves both MW freeze-drying and pulse-spouted bed drying. While the former allows drying time and energy consumption to be reduced, the spouted bed drying system compensates limitation as annulus aeration and slow solids turnover of conventional spouted bed. More highly transparent gels, higher foam stability, and higher emulsifying indices have been found in egg white powders prepared using this system (Wang, Zhang, Adhikari, Mujumdar, & Zhou, 2013). The authors reported that shorter drying times and lower product temperatures were able to minimize egg white protein denaturation and loss when compared to the conventional freeze-drying technique.

It has been reported that better total sugar content preservation has been achieved by combining vacuum MW with spouted bed drying in a carrot drying process. The benefits were reportedly caused by higher dehydration rates, shorter drying times and the use of lower temperatures (Oliveira et al., 2016).

6.2.2 Microwave Freezing

Freezing is a long-established food preservation process that provides foods with a long storage life and high nutritional quality. We must, however, consider the fact that the formation of ice crystals during treatment can often cause undesirable physical changes to food structure. Fast freezing and the formation of small ice crystals can offer some advantages in terms of quality, reduction in energy consumption, and yields improvement. MW-assisted freezing is still in an early research and development phase. Although the real advantages of this technology are not yet fully clear, MW radiation may reduce damage to meat tissue and consequently provide frozen meat with a better texture (Xanthakis, Havet, Chevallier, Abadie, & Le-Bail, 2013). Although the oscillated temperature decreases caused by the use of MW during cooling lead to longer freezing times, the average ice crystal size decreases when meat samples are frozen under MW field, as compared to the conventional freezing process. This is probably due to the limited temperature oscillation that occurs during the genesis of ice nuclei and crystal growth.

6.2.3 Microwave Heating and Cooking

MW heating is based on volumetric heating, which causes foods to heat instantaneously. MW electromagnetic fields induce dipole rotation in foods and the consequent friction between molecules generates heat. Electromagnetic waves penetrate the food through its surface and spread inside, while energy is absorbed and transformed into thermal energy. Lowering the water content of a food may result in the reduction of MW absorption and an increase in MW penetration depth. Salt content must also be assessed, as increasing it corresponds to an improvement in dielectric loss, therefore reducing MW penetration. Fat can also play a major role because its thermal properties lead to an increase in the MW heating rate and heating uniformity, allowing higher maximum temperatures to be reached (Stratakos & Koidis, 2015). Furthermore, the presence of bones in meat affects dielectric heating as they can shield it from MW in some cases, thus resulting in reduced quality and safety. The presence of the free ions contained in additives can also influence the dielectric properties of foods (Marra, Zhang, & Lyng, 2009). The problem of microwave heating uniformity may be managed by using a feedback control loop system in combination with a thermo-camera to monitor temperature distribution (Guo et al., 2017).

MW heating can affect the antioxidant activity of bioactive components and the anti-nutritional effects (i.e. trypsin inhibition) of haemagglutinin activity, tannins, saponins, and phytate. No crust is formed when meat is cooked with MW and total cooking loss is higher than with other common cooking methods, such as grilling, roasting or braising (Domínguez, Gómez, Fonseca, & Lorenzo, 2014). On the other hand, vegetables processed with MW maintain higher bioactive component contents than those processed by other cooking methods due to the shorter heating time. In fact, antioxidant activity and bioactive components were retained better when cooking was carried out both with and without a small amount of water. However, a significant reduction in polyphenol content was found when vegetables, such as kale, green beans, and tomato, underwent MW cooking with water (Dolinsky et al., 2016), as it led to the softening and rupturing of the lignocellulosic structure and soluble bioactive compounds were released from the food matrix. The effectiveness of MW cooking in reducing anti-nutritional factors and increasing protein digestibility in foods has been demonstrated in lentils (Hefnawy, 2011). After MW cooking, a 93.3% drop in the contents of trypsin inhibitor, a 34.4% drop in tannins and a 39.2% drop in phytate content was found, while an improved protein efficiency ratio and *in vitro* protein digestibility were reported for these legumes. While MW heating can lead to changes in the sensory attributes, such as texture and color properties, of food products, it must be stressed that both food type and heating conditions can significantly influence the final effects, which can vary wildly. For example, whereas tougher texture and a higher shrinkage rate can often occur when red meat is cooked by MW, as compared to roasting with traditional heating (Jouquand et al., 2015), the hardness of chicken steak can actually be lower than when it is boiled or grilled (Choi et al., 2016). A significant decrease in shear force is found after microwaving potato tubers as the cohesive forces between cells decrease. Furthermore, protein denaturation can affect food color, such as when the thermal denaturation of myoglobin and other proteins causes a reduction in redness in meat. The exposure time of foods is generally reduced by an increase in MW power, thus limiting the denaturation of proteins and lowering the effects of color change. Lowered lightness, higher redness, and lowered yellowness can all be achieved by increasing MW power levels (Guo et al., 2017).

While frying is a means of processing foods that, on the one hand, is appreciated because it imparts flavor, taste, color, and crispness, on the other, the danger of excessive consumption of fats is well known. However, pre-treatment techniques have been successfully applied, prior to frying, to reduce oil uptake and improve the quality parameters of fried foods. MW technology is one of the recent and novel processes to be used for this purpose and to improve the quality of fried foods. Two main paths can be pursued to this aim; either the frying process can be directly performed inside the MW device, which generates the heat for frying, or the food material can be pre-cooked using MW before deep fat frying is carried out. A comparison between the MW frying and conventional frying of chicken breast meat at 180 °C showed that MW frying required shorter processing times and generated a higher heat transfer coefficient, which was probably due to the higher turbulence experienced during MW frying that led to lower oil absorption by food (Sensoy, Sahin, & Sumnu, 2013). MW-fried potato strips (400 W for 1 min) showed an acrylamide content reduction of 88% when compared to samples that were conventionally fried at 170 °C for 4.5 min (Sahin, Sumnu, & Oztop, 2007). Su, Zhang, Zhang, Adhikari, and Yang (2016) have observed that potato chips fried by MW-assisted vacuum frying (vacuum degree of 0.065, 0.075 and 0.085 MPa) offered better quality in terms of lower oil content, faster moisture evaporation rates, crispier chips and less color change, as compared to conventional vacuum frying. The significant effects on moisture loss and oil uptake were related to the MW pre-treatment of frozen coated chicken nuggets during deep-fat frying (Ngadi, Wang, Adedeji, & Raghavan, 2009). Interactions with polar molecules and charged food particles mean that MW generate heat via a mechanism that is different to that of conventional heating and that shows advantages in terms of energy and time savings and the improved nutritional quality of some foods. However, MW baking does entail some problems including a lack of volume, tough or firm textures, a lack of browning and flavor development, in cakes and bread baking, for example. One must also consider that the physicochemical changes and interactions with major ingredients that normally occur over a lengthy baking period cannot always be completed under MW (Sumnu, 2001). The problem of low volume is a result of the incomplete gelatinization of starch. The use of pre-gelatinized starches or starches with high dielectric properties and low gelatinization enthalpy may be a means to combat this issue. Hydrocolloids can limit the high moisture loss in MW-baked cakes, while the problem of the firmness and toughness of baked products can be resolved by the use of additives. An increase in the surface temperature of the dough is helpful for the browning reaction; susceptors and commercial coatings can be used for this purpose. Nevertheless, improving the quality of MW-baked products is still a challenging topic for the food industry.

Some further examples of MW use in food preparation can be found in doughnut proofing and frying, the drying of pasta products and meat tempering. When tempering food, it is most important that the product enters the MW tunnel at a uniform temperature, without any incipient thawing. Moreover, the puffing and drying of snack foods is another widespread application that can boast of products being specifically developed for MW processing. One of the advantages of this technology is the treatment of packaged food due to the use of plastic packaging materials that are transparent to MW radiation.

6.2.4 Microwave Sterilization

During sterilization, all vegetative microorganisms and their spores are inactivated via the application of physical and chemical processes. The food industry commonly ensures food safety using thermal processing, which can affect organoleptic properties, nutritional values, and texture, while also generating by-products. The growing attention that consumers pay to food and health has driven the attention of the food industry and scientists towards electromagnetic field-based, non-thermal processing technologies that can improve food shelf-life, and have a minimal impact on nutritional value and sensory characteristics (Pan, Sun, & Han, 2017). Selective heating, electroporation, cell membrane rupture, and magnetic field coupling are among several mechanisms that are related to MW pasteurization and sterilization. The destruction of pathogens can occur when the temperature of microbial bodies is higher than that of the surrounding fluid, as occurs during MW heating. Furthermore, cellular material can leak from cells after the opening of their pores by an electrical potential applied across the cell membrane (electroporation phenomenon), while magnetic field coupling can denature proteins and break nucleic acids (Guo et al., 2017). Sterilization effectiveness may be correlated to the power applied and the temperature-increasing radiation dose, depending on the type of food. While the thermal effect is obviously well understood, it is believed that non-thermal effects result from direct interaction between the electric field and specific molecules, in a way that is not related to a macroscopic temperature effect. The existence of some non-thermal MW effects has been proven by the reversible leakage of cellular cytosolic fluids and a visual representation of the morphological changes that occur in *E. coli* cells (Shamis et al., 2011), MW-assisted pasteurization has been effectively applied to both fluid and solid food materials, such as fresh juices, milk and in-shell eggs (Chandrasekaran, Ramanathan, & Basak, 2013). The short exposure time means that MW sterilization has a mild effect on bioactive substances and antioxidants, while causing enzyme inactivation, thus maintaining the quality of the product. Furthermore, the textural and color properties of food were generally less influenced by MW sterilization, as compared to conventional methods, but lengthy treatment could still affect these properties. Packaging material components might degrade and migrate into food materials during the sterilization of packed food products and the degradation of polymer chains, additives or adhesive layers can produce unwanted by-products during high-temperature MW heating. Therefore, glass, paper, and ceramics are the preferred packaging types for MW treatment (Guillard, Mauricio-Iglesias, & Gontard, 2010).

In conclusion, MW sterilization has the capacity to completely inactivate microorganisms and effectively destroy enzyme activity, while only having a mild effect on the antioxidant activity, texture, and color of food products, as compared to conventional pasteurization.

6.3 Flavoring and Food Component Extraction

MW strongly promotes extraction from vegetal matrices, both when used alone and when combined with other techniques. Microwave-assisted extraction (MAE) has gained enormous popularity as a preferred method for the recovery of various active compounds from food materials. As food flavorings strongly influence food evaluation and consumer satisfaction, they have always received a lot of attention from the food industry, although it is only over recent decades that enabling technologies have revolutionized classic extraction procedures. The MW heating of fresh plant material is a simple way to achieve the direct distillation of essential oils (Lucchesi, Chemat, & Smadja, 2004), while the use of MAE in the large-scale recovery of food components should be carried out to translate laboratory procedures into industrial applications.

Among the major benefits that MAE can offer, we find shorter extraction times and the preservation of thermally susceptible compounds (e.g. natural antioxidants) as well as the fact that it is carried out under environmentally friendly conditions. MAE's superiority over conventional methods, when it comes to extracting food target components, is principally related to its volumetric heating and high penetration power. The process works by heating the moisture inside cells, which evaporates and produces high pressure within the cell wall. The consequent pressure that grows inside the biomaterial modifies its physical properties and increases its porosity. It follows that extracting solvents thus penetrate matrices, providing improved compound yields and significant decreases in extraction times and solvent requirements in an environmentally benign approach. Moreover, MAE can be operated both under pressure and at atmospheric pressure (Gil-Chávez et al., 2013).

Solvent choice is one of the most important factors to consider here; higher dielectric constants generally correspond to a higher capacity to absorb MW energy, which will then be heated faster than the matrix being extracted. Solvent properties and selectivity can be modified by combining different solvents. In the case of thermolabile compounds, a solvent system with relatively lower dielectric properties can be recommended as it will ensure that the solvent temperature will remain lower, thus cooling off solutes once they are liberated into the solvent. Obviously, solvent polarity must be evaluated according to the type of compound to be extracted; less polar solvents can be used for the extraction of flavonoid aglycones and more polar solvents are used for extraction of flavonoid glycosides and anthocyanin. Although extractant choice has generally been between organic solvents or water, various hydrotropic liquids, two-phase and micellar solutions can nowadays be used to minimize solvent usage and toxicity, waste production and energy consumption, thus providing a greener implementation strategy (Ekezie, Sun, & Cheng, 2017).

While extraction yields are certainly related to extraction time, the prolonged application of MW may lead to target compound degradation if the solute/solvent system is overheated. A multiple step method that provides consecutive extraction cycles can also be used to improve extraction yields and prevent prolonged heating in the same volume of solvent. The saturation of the solvent with the solute is avoided by the use of fresh batches of solvent, thus increasing mass transfer and extraction kinetics. MW power, temperature, moisture content, sample matrix particle size, and the solid/liquid ratio can all affect extraction kinetics, and the right conditions must be adapted for specific matrices if yields are to be maximized and product quality preserved.

In order to further promote extraction efficacy, MW systems can be synergistically coupled with other technologies, such as ultrasound, supercritical or subcritical fluid extractors, enzyme-assisted extraction, hydrodiffusion and so on.

Consumer demand for safe and minimally processed foods with high-quality attributes have stimulated the search to find innovative food-processing techniques that can limit changes in food products. For example, MW heating can be used in fruit-juice extraction. In particular, MW hydrodiffusion has been used in plum, apricot and grape juice extraction processes giving rapid extractions from fresh and frozen samples. The highest yields were obtained from frozen fruit and at low power, producing juices characterized by very bright colors, high acidity in plums and apricots and fresh fruit flavors (Cendres, Chemat, Maingonnat, & Renard, 2011). Polysaccharide extraction process enhancement has been obtained in a number of matrices, including apple and grape pomace, wheat bran and dragon fruit peel (Tejada-Ortigoza, Garcia-Amezquita, Serna-Saldívar, & Welti-Chanes, 2016).

Solvent-free MW extraction is a green technology, which works thanks to a combination of MW heating and dry distillation, performed at atmospheric pressure and without the addition of solvents or water, is a good alternative technique for the extraction of essential oils from aromatic plants and spices. Furthermore, the fast extraction of aromatic and non-volatile compounds from spices and aromatic plants (e.g. terpenes, capsaicinoids, paprika red pigment, piperine, curcumin etc.) can be achieved using MW-assisted techniques (Kubra et al., 2016). MAE is also gaining interest as an alternative means for the extraction of both water- and oil-based bioactive plant pigments, such as β-carotene from carrots, aloin A from *Aloe vera* and curcuminoids from turmeric (Ngamwonglumlert, Devahastin, & Chiewchan, 2017).

6.4 Combined Hybrid MW Technologies

Combined MW and ultrasound (e.g. drying, extraction, enzyme activity, etc.), MW-assisted infrared heating, MW-assisted osmotic dehydration, MW-powered plasma, combined MW-Ohmic heating can all offer great advantages (Cravotto et al., 2008). MW-assisted food processing technologies are a hybrid combination that can enhance both conventional and non-conventional food processes with MW radiation and overcome the shortcomings of traditional techniques. The advantages offered to translate into energy conservation, better product quality, reduced time and lower operational costs for food processing (Chizoba Ekezie et al., 2017).

Adding MW energy to vacuum processing helps to preserve important sensory features, including flavor, appearance, and texture, due to the absence of air, and also furnishes higher production output and lower running costs. Vacuum drying and vacuum frying can be effectively assisted by MW heating. In the first case, the principal effect of this coupling is the rapid evaporation of moisture, which is also useful in enhancing the survival rate and metabolic activity of probiotic and starter cultures during dehydration (Ambros, Bauer, Shylkina, Foerst, & Kulozik, 2016). Vacuum frying lowers the boiling point of frying oil and the moisture in fried foods, which reduces acrylamide content and adverse oil quality effects while preserving natural color and flavors. The combination of MW vacuum frying, after either pre-frying or pulsed-spouted MW vacuum drying followed by vacuum pre-frying, enhances the quality of fried products and gives texture and color attributes that are comparable to those given by traditional methods (Quan et al., 2016; Stratakos & Koidis, 2015).

The coupling of convective drying and MW radiation is a feasible hybrid technology, but is best carried out in a pulsed manner, namely by applying MW at pulsed rates (intermittent MW-convective drying), because product quality can be affected by the overheating caused by continuous MW radiation (Kumar, Joardder, Farrell, Millar, & Karim, 2016).

Freeze-drying is a method that is commonly used for the removal of moisture from heat-sensitive foods, but it can face the drawbacks of long drying times, low productivity and high-energy costs. MW-assisted freeze-drying can lead to better process efficiency because of its shorter processing times, rapid energy dissipation throughout materials, high-energy savings, and retention of volatiles (Duan, Liu, Ren, Liu, & Liu, 2016). During freezing, large ice crystals can generate irreversible tissue damage as they induce mechanical and biochemical stress on the structure of food materials. MW-energy assisted freezing can reduce the size of the ice crystals formed and reduce microstructure damage in food products (Xanthakis, Le-Bail, & Ramaswamy, 2014).

MW energy can be combined with osmotic dehydration to generate rapid and uniform heating and optimize drying times while making changes in dielectric characteristics can lead to enhancements in solute uptake. Moreover, high food quality and low moisture content can be obtained using pulsed-MW-vacuum osmotic drying (Patel & Sutar, 2016).

Emerging food processing technologies can also be assisted by MW. Infrared baking, drying, roasting, and tempering are technologies with weak penetrating power and thus only generate surface heating. The greater penetration depth of MW provides volumetric heating and minimal temperature differences between the surface and the interior of food materials; properties that can prevent food swelling and the fracturing generated by prolonged exposure to infrared radiation. Combining food browning and crust formation, produced by infrared heating on the one hand, with time savings and efficient temperature distribution inside the product, furnished by MW on the other, can significantly improve the quality of bakery products (Ozge, Sumnu, & Meda, 2006). Moreover, MW-infrared heating can be useful in protecting the desirable properties of food products after tempering treatment, while also optimizing process times (Seyhun, Ramaswamy, Sumnu, Sahin, & Ahmed, 2009). The moisture loss and firmness of food products baked using this coupled technique were found to improve with increases in the power of both systems, although increasing MW power was more effective due to high internal pressure and concentration gradients.

The advantages provided to food processing by the use of ultrasound are well known nowadays. However, the physicochemical effects of ultrasonication can affect product quality parameters, such as adding off-flavors, changing physical parameters and causing degradation phenomena. The combined MW and ultrasound-assisted extraction technique is an efficient extraction method that has previously been described (Chemat & Cravotto, 2011). MW energy destabilizes weak hydrogen bonds by enhancing the rotation of molecular dipoles and this, together with dissolved ion movement, increases solvent penetration into the matrix and solvation. Ultrasound, however, enhances extraction efficiency, through cavitation, mechanical function, and thermal effects, by disrupting cell walls and facilitating solvent access to cell content. Moreover, matrix fragmentation can improve the hydration process without leading to any considerable chemical degradation. The MW-ultrasound hybrid technique can also enhance drying processes and reduce the risk of oxidation and degradation. The volumetric heating of the vapors generated inside food material leads to the development of an internal pressure gradient that forces the water outside, while product shrinkage is prevented by using shorter drying times and lower temperatures than traditional methods (Kowalski et al., 2016).

The use of MW-ultrasound to assist enzyme-based reactions is a new approach that has shown the potential to increase the efficiency and specificity of enzymes during reactions, including in starch and dextran hydrolysis (Bashari, Jin, Wang, & Zhan, 2016).

Microbial decontamination is among the applications of cold plasma in the food industry. This non-thermal technology is able to damage DNA and inactivate microbes via the reactive species that it produces. Within this context, MW can enhance decontamination processes by inducing plasma formation and increasing electron density. For example, microbial spores were found to be more easily attacked by excited molecules due to the cleavage of disulfide bonds in the protein coat of their cells (Kim, Oh, Won, Lee, & Min, 2017). Sterilization can also be obtained using electronic pasteurization, which makes use of high-energy electrons. While this electron-beam irradiation is able to damage the DNA and RNA chains of pathogens, the high electron dose required can have a negative effect on food. Synergistic action is ascribed to the combination of this kind of irradiation with MW energy volumetric heating and this action may be able to alter the sensitivity of microorganisms to electrons and enhance food shelf life (Mulmule et al., 2017). Finally, enhanced food particle heating uniformity has been observed when ohmic and MW heating was combined and the problems caused by the differing electrical conductivities of various food components were limited (Choi, Nguyen, Lee, & Jun, 2011).

6.5 Conclusions

MW energy has proven itself to be one of the best modern techniques with which to enhance current industrial food processing and minimize changes to food quality, while also comparing favorably to traditional methods. The most successful applications, thawing/tempering, blanching, roasting, frying, and freezing, generally display improved nutritional and sensory profile retention. Energy savings, higher extraction efficiency, and a greener approach, due to reduced solvent use, are foremost among the many advantages that the technique offers, while MW drying, heating, and sterilization play a significant role in food quality and safety control. The main limitation of the technique is the high investment cost required in the design and assembly of dedicated MW systems, on both large and pilot scales. However, other emerging food processing technologies, such as high-pressure processing, radio frequency heating and ultraviolet light, can be combined with MW energy, making further investigations into these potential hybrid technologies, their feasibility and effects on food interesting prospects for the future. Despite the advantages provided by MW heating, its non-uniform temperature distribution is its major disadvantage, as this can lead to hot and cold spots forming in foods. Keeping the heated material in constant motion and the proper selection of packaging materials are among the proposed solutions, which aim to optimize processing parameters, ensure uniform MW power absorption and therefore prevent the formation of cold/hot regions. Recent research into MW equipment design, MW/material interactions, dielectric property measurement, and materials processing continues to expand interest in MW techniques.

References

Alibas, I. (2007). Energy consumption and colour characteristics of nettle leaves during microwave, vacuum and convective drying. *Biosystems Engineering, 96*(4), 495–502.

Ambros, S., Bauer, S. a. W., Shylkina, L., Foerst, P., & Kulozik, U. (2016). Microwave-vacuum drying of lactic acid bacteria: Influence of process parameters on survival and acidification activity. *Food and Bioprocess Technology, 9*(11), 1901–1911.

Arikan, M. F., Ayhan, Z., Soysal, Y., & Esturk, O. (2012). Drying characteristics and quality parameters of microwave-dried grated carrots. *Food and Bioprocess Technology, 5*(8), 3217–3229.

Bashari, M., Jin, Z., Wang, J., & Zhan, X. (2016). A novel technique to improve the biodegradation efficiency of dextranase enzyme using the synergistic effects of ultrasound combined with microwave shock. *Innovative Food Science & Emerging Technologies, 35*, 125–132.

Buffler, C. R. (1993). *Microwave cooking and processing: Engineering fundamentals for the food scientist*. New York, NY: Van Nostrand Reinhold.

Cendres, A., Chemat, F., Maingonnat, J.-F., & Renard, C. M. G. C. (2011). An innovative process for extraction of fruit juice using microwave heating. *LWT – Food Science and Technology, 44*(4), 1035–1041.

Chandrasekaran, S., Ramanathan, S., & Basak, T. (2013). Microwave food processing – A review. *Food Research International, 52*(1), 243–261.

Chemat, F., & Cravotto, G. (2011). Combined methods of assisted extraction. In N. Lebovka, E. Vorobiev, & F. Chemat (Eds.), *Enhancing extraction processes in the food industry* (pp. 173–193). Taylor & Francis/CRC Press, Boca Raton, Florida, US.

Chizoba Ekezie, F.-G., Sun, D.-W., Han, Z., & Cheng, J.-H. (2017). Microwave-assisted food processing technologies for enhancing product quality and process efficiency: A review of recent developments. *Trends in Food Science & Technology, 67*, 58–69.

Choi, W., Nguyen, L. T., Lee, S. H., & Jun, S. (2011). A microwave and ohmic combination heater for uniform heating of liquid–particle food mixtures. *Journal of Food Science, 76*(9), E576–E585.

Choi, Y.-S., Hwang, K.-E., Jeong, T.-J., Kim, Y.-B., Jeon, K.-H., Kim, E.-M., … Kim, C.-J. (2016). Comparative study on the effects of boiling, steaming, grilling, microwaving and superheated steaming on quality characteristicsof marinated chicken steak. *Korean Journal for Food Science of Animal Resources, 36*(1), 1–7.

Cravotto, G., Boffa, L., Mantegna, S., Perego, P., Avogadro, M., & Cintas, P. (2008). Improved extraction of vegetable oils under high-intensity ultrasound and/or microwaves. *Ultrasonics Sonochemistry, 15*(5), 898–902.

Cravotto, G., & Carnaroglio, D. (2017). *Microwave chemistry*. Berlin,Germany/Boston, MA: De Gruyter.

Dehnad, D., Jafari, S. M., & Afrasiabi, M. (2016). Influence of drying on functional properties of food biopolymers: From traditional to novel dehydration techniques. *Trends in Food Science & Technology, 57*, 116–131.

Dolinsky, M., Agostinho, C., Ribeiro, D., Rocha, G. D. S., Barroso, S. G., Ferreira, D., … Fialho, E. (2016). Effect of different cooking methods on the polyphenol concentration and antioxidant capacity of selected vegetables. *Journal of Culinary Science & Technology, 14*(1), 1–12.

Domínguez, R., Gómez, M., Fonseca, S., & Lorenzo, J. M. (2014). Effect of different cooking methods on lipid oxidation and formation of volatile compounds in foal meat. *Meat Science, 97*(2), 223–230.

Duan, X., Liu, W. C., Ren, G. Y., Liu, L. L., & Liu, Y. H. (2016). Browning behavior of button mushrooms during microwave freeze-drying. *Drying Technology, 34*(11), 1373–1379.

Ekezie, F.-G. C., Sun, D.-W., & Cheng, J.-H. (2017). Acceleration of microwave-assisted extraction processes of food components by integrating technologies and applying emerging solvents: A review of latest developments. *Trends in Food Science & Technology, 67*, 160–172.

Gensback, H. (1933). Cooking by ultra short waves. *Short Wave Craft-the Radio Experimenter's Magazine*, 394–429.

Gil-Chávez, G. J., Villa, J. A., Ayala-Zavala, J. F., Heredia, J. B., Sepulveda, D., Yahia, E. M., & González-Aguilar, G. A. (2013). Technologies for extraction and production of bioactive compounds to be used as nutraceuticals and food ingredients: An overview. *Comprehensive Reviews in Food Science and Food Safety, 12*(1), 5–23.

Guillard, V., Mauricio-Iglesias, M., & Gontard, N. (2010). Effect of novel food processing methods on packaging: Structure, composition, and migration properties. *Critical Reviews in Food Science and Nutrition, 50*(10), 969–988.

Guo, Q., Sun, D.-W., Cheng, J.-H., & Han, Z. (2017). Microwave processing techniques and their recent applications in the food industry. *Trends in Food Science & Technology, 67*, 236–247.

Hefnawy, T. H. (2011). Effect of processing methods on nutritional composition and anti-nutritional factors in lentils (*Lens culinaris*). *Annals of Agricultural Sciences, 56*(2), 57–61.

Jouquand, C., Tessier, F. J., Bernard, J., Marier, D., Woodward, K., Jacolot, P., … Laguerre, J.-C. (2015). Optimization of microwave cooking of beef burgundy in terms of nutritional and organoleptic properties. *LWT – Food Science and Technology, 60*(1), 271–276.

Kim, J. E., Oh, Y. J., Won, M. Y., Lee, K.-S., & Min, S. C. (2017). Microbial decontamination of onion powder using microwave-powered cold plasma treatments. *Food Microbiology, 62*, 112–123.

Kowalski, S. J., Pawłowski, A., Szadzińska, J., Łechtańska, J., & Stasiak, M. (2016). High power airborne ultrasound assist in combined drying of raspberries. *Innovative Food Science & Emerging Technologies, 34*, 225–233.

Kubra, I. R., Kumar, D., & Rao, L. J. M. (2016). Emerging trends in microwave processing of spices and herbs. *Critical Reviews in Food Science and Nutrition, 56*(13), 2160–2173.

Kumar, C., Joardder, M. U. H., Farrell, T. W., Millar, G. J., & Karim, M. A. (2016). Mathematical model for intermittent microwave convective drying of food materials. *Drying Technology, 34*(8), 962–973.

Li, F., Chen, G., Zhang, B., & Fu, X. (2017). Current applications and new opportunities for the thermal and non-thermal processing technologies to generate berry product or extracts with high nutraceutical contents. *Food Research International, 100*, 19–30.

Li, Z. Y., Wang, R. F., & Kudra, T. (2011). Uniformity issue in microwave drying. *Drying Technology, 29*(6), 652–660.

Lucchesi, M. E., Chemat, F., & Smadja, J. (2004). Solvent-free microwave extraction of essential oil from aromatic herbs: Comparison with conventional hydro-distillation. *Journal of Chromatography A, 1043*(2), 323–327.

Marra, F., Zhang, L., & Lyng, J. G. (2009). Radio frequency treatment of foods: Review of recent advances. *Journal of Food Engineering, 91*(4), 497–508.

Mullin, J. (1994). MW processing. In G. W. Gould (Ed.), *New methods of food preservation* (pp. 112–134). Springer New York City, US.

Mulmule, M. D., Shimmy, S. M., Bambole, V., Jamdar, S. N., Rawat, K. P., & Sarma, K. S. S. (2017). Combination of electron beam irradiation and thermal treatment to enhance the shelf-life of traditional Indian fermented food (Idli). *Radiation Physics and Chemistry, 131*, 95–99.

Ngadi, M. O., Wang, Y., Adedeji, A. A., & Raghavan, G. S. V. (2009). Effect of microwave pretreatment on mass transfer during deep-fat frying of chicken nugget. *LWT – Food Science and Technology, 42*(1), 438–440.

Ngamwonglumlert, L., Devahastin, S., & Chiewchan, N. (2017). Natural colorants: Pigment stability and extraction yield enhancement via utilization of appropriate pretreatment and extraction methods. *Critical Reviews in Food Science and Nutrition, 57*(15), 3243–3259.

Nijhuis, H. H., Torringa, H. M., Muresan, S., Yuksel, D., Leguijt, C., & Kloek, W. (1998). Approaches to improving the quality of dried fruit and vegetables. *Trends in Food Science & Technology, 9*(1), 13–20.

Oliveira, S. M., Brandão, T. R. S., & Silva, C. L. M. (2016). Influence of drying processes and pretreatments on nutritional and bioactive characteristics of dried vegetables: A review. *Food Engineering Reviews, 8*(2), 134–163.

Ozge, S., Sumnu, G., & Meda, V. (2006). The effect of different formulations on physical properties of cakes baked with microwave and near infrared-microwave combinations. *Journal of Microwave Power and Electromagnetic Energy, 41*(1), 20–26.

Pan, Y., Sun, D.-W., & Han, Z. (2017). Applications of electromagnetic fields for nonthermal inactivation of microorganisms in foods: An overview. *Trends in Food Science & Technology, 64*, 13–22.

Patel, J. H., & Sutar, P. P. (2016). Acceleration of mass transfer rates in osmotic dehydration of elephant foot yam (*Amorphophallus paeoniifolius*) applying pulsed-microwave-vacuum. *Innovative Food Science & Emerging Technologies, 36*, 201–211.

Puligundla, P., Abdullah, S. A., Choi, W., Jun, S., Oh, S.-E., & Ko, S. (2013). Potentials of Microwave Heating technology for select food processing applications - a brief overview and update. *Journal of Food Processing & Technology, 4*(11), 1–9.

Quan, X., Zhang, M., Fang, Z., Liu, H., Shen, Q., & Gao, Z. (2016). Low oil French fries produced by combined pre-frying and pulsed-spouted microwave vacuum drying method. *Food and Bioproducts Processing, 99*, 109–115.

Sahin, S., Sumnu, G., & Oztop, M. H. (2007). Effect of osmotic pretreatment and microwave frying on acrylamide formation in potato strips. *Journal of the Science of Food and Agriculture, 87*(15), 2830–2836.

Sensoy, I., Sahin, S., & Sumnu, G. (2013). Microwave frying compared with conventional frying via numerical simulation. *Food and Bioprocess Technology, 6*(6), 1414–1419.

Seyhun, N., Ramaswamy, H., Sumnu, G., Sahin, S., & Ahmed, J. (2009). Comparison and modeling of microwave tempering and infrared assisted microwave tempering of frozen potato puree. *Journal of Food Engineering, 92*(3), 339–344.

Shamis, Y., Taube, A., Mitik-Dineva, N., Croft, R., Crawford, R. J., & Ivanova, E. P. (2011). Specific electromagnetic effects of microwave radiation on *Escherichia coli*. *Applied and Environmental Microbiology, 77*(9), 3017–3022.

Sorour, H., & El-Mesery, H. (2014). Effect of microwave and infrared radiation on drying of onion slices. *International Journal of Research in Applied Natural and Social Science, 2*, 119–130.

Stratakos, A. C., & Koidis, A. (2015). Suitability, efficiency and microbiological safety of novel physical technologies for the processing of ready-to-eat meals, meats and pumpable products. *International Journal of Food Science & Technology, 50*(6), 1283–1302.

Su, Y., Zhang, M., Zhang, W., Adhikari, B., & Yang, Z. (2016). Application of novel microwave-assisted vacuum frying to reduce the oil uptake and improve the quality of potato chips. *LWT, 73*, 490–497.

Sumnu, G. (2001). A review on microwave baking of foods. *International Journal of Food Science & Technology, 36*(2), 117–127.

Tejada-Ortigoza, V., Garcia-Amezquita, L. E., Serna-Saldívar, S. O., & Welti-Chanes, J. (2016). Advances in the functional characterization and extraction processes of dietary fiber. *Food Engineering Reviews, 8*(3), 251–271.

Vadivambal, R., & Jayas, D. S. (2007). Changes in quality of microwave-treated agricultural products - a review. *Biosystems Engineering, 98*(1), 1–16.

Wang, Y., Zhang, M., Adhikari, B., Mujumdar, A. S., & Zhou, B. (2013). The application of ultrasound pretreatment and pulse-spouted bed microwave freeze drying to produce desalted duck egg white powders. *Drying Technology, 31*(15), 1826–1836.

Witkiewicz, K., & Nastaj, J. F. (2010). Simulation strategies in mathematical modeling of microwave heating in freeze-drying process. *Drying Technology, 28*(8), 1001–1012.

Xanthakis, E., Havet, M., Chevallier, S., Abadie, J., & Le-Bail, A. (2013). Effect of static electric field on ice crystal size reduction during freezing of pork meat. *Innovative Food Science & Emerging Technologies, 20*, 115–120.

Xanthakis, E., Le-Bail, A., & Ramaswamy, H. (2014). Development of an innovative microwave assisted food freezing process. *Innovative Food Science & Emerging Technologies, 26*, 176–181.

Chapter 7
Cold Plasma Effects on the Nutritional, Textural and Sensory Characteristics of Fruits and Vegetables, Meat, and Dairy Products

George Amponsah Annor

7.1 Introduction

The non-thermal nature of cold plasma processing has brought it to the spotlight in recent times as an alternative food processing technology, especially for foods sensitive to heat. Simply defined as the generation of short-lived reactive species by the application of electricity to gas, non-thermal plasma has become an important food processing technology. Figure 7.1 shows a schematic presentation of atmospheric cold plasma processing of food products. Depending on the plasma technique used (i.e. corona discharge, dielectric barrier discharge, gliding arch, plasma jets, and radio frequency discharges (Scholtz, Pazlarova, Souskova, Khun, & Julak, 2015)), different reactive species are produced, usually from vibrationally and electronically excited nitrogen and oxygen. The type of reactive species produced largely depends on the type of gas used. The gases mostly used are but not limited to oxygen, nitrogen, argon, hydrogen, air and their mixtures. These reactive species react with surfaces, they come into contact with resulting in modifications. The effects of cold plasma on the various food components such as proteins (Bahrami et al., 2016; Dong, Gao, Xu, & Chen, 2017; Misra et al., 2015; Takai et al., 2014; Yasuda, Miura, Kurita, Takashima, & Mizuno, 2010), starch (Bastos, Santos, & Simao, 2014; Kim & Min, 2017; Thirumdas, Trimukhe, Deshmukh, & Annapure, 2017), lipids (Bahrami et al., 2016; Sarangapani, Ryan Keogh, Dunne, Bourke, & Cullen, 2017), and phenolics (Amini & Ghoranneviss, 2016; Grzegorzewski, Ehlbeck, Schlüter, Kroh, & Rohn, 2011) have been previously reported. One of the main applications of cold plasma in food processing is for the sterilization of food products (Los, Ziuzina, Boehm, Cullen, & Bourke, 2017; Misra & Jo, 2017; Selcuk, Oksuz, & Basaran, 2008). Other applications such as food quality improvement, packaging applications, surface modifications, and the degradation of toxins in foods have

G. A. Annor (✉)
Department of Food Science and Nutrition, University of Minnesota, St. Paul, MN, USA
e-mail: gannor@umn.edu

S. Roohinejad et al. (eds.), *Effect of Emerging Processing Methods on the Food Quality*,
https://doi.org/10.1007/978-3-030-18191-8_7

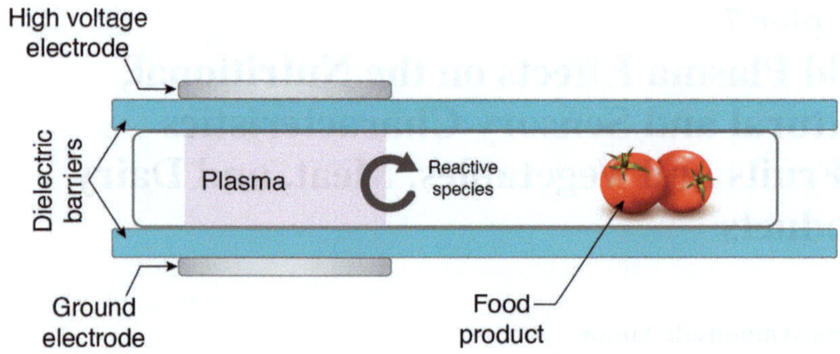

Fig. 7.1 Schematic presentation of atmospheric cold plasma processing of food products

been reported (Akocak, 2016; Han et al., 2016; Shi, Cooper, Stroshine, Ileleji, & Keener, 2017). In this chapter, the advantages and challenges of using cold plasma on the quality of fruits, vegetables, meat and dairy products are highlighted. The discussion is focused on the effects of cold plasma on the nutritional, textural and sensory properties of fruits and vegetables, meat and dairy products.

7.2 Advantageous and Challenges of Using Cold Plasma on Food Quality

7.2.1 Nutritional Properties

7.2.1.1 Fruits and Vegetables

The reactive species produced by cold plasma may affect the bioactive compounds in fruits and vegetables leading to changes in their nutritional composition and functional properties. Previous studies have reported that the application of this technology might affect the stability of some nutrients. For instance, Bursać Kovačević et al. (2016) investigated the effect of cold plasma on the stability of anthocyanins in pomegranate juice. In this study, the application of different treatment times, juice volume and gas flow were evaluated. The authors observed an increase in anthocyanin content after treating the juice with cold plasma. The increase in anthocyanin contents in the juice was attributed to the disruption of the fruit cell walls. The application of this technology not only increased the anthocyanins' content in the juice but also improved the stability.

Elez Garofulić et al. (2015) have reported the degradation of anthocyanins and phenolic compounds treated with cold plasma for longer periods of time. The degradation of anthocyanins and phenolic compounds were observed in the sour cherry Marasca juice samples treated with cold plasma. The extent of degradation positively correlated with an increase in plasma exposure. The degradation of anthocyanins and phenolic compounds, when exposed to plasma for extended periods of time, could be attributed to the reaction of these compounds with the reactive radicals or oxygen spices such as hydroxyl radicals, peroxyl radicals, atomic and singlet oxygen (Brandenburg et al., 2007).

In another study, Grzegorzewski et al. (2011) investigated the effect of cold plasma treatment on the phenolic profile of lamb's lettuce. The authors observed different degradation rates in various phenolic compounds. All phenolic acids in the fresh lettuce leaves decreased by increasing the plasma exposure. The rate of degradation of phenolic acids, when exposed to plasma, was slower than flavonoids. Luteolin was unaffected by plasma exposure, whereas a significant increase was observed for diosmetin. A significant decrease in the anthocyanin content of blueberries was observed after 90 s atmospheric cold plasma treatment.

A few studies also reported the effect of cold plasma on the ascorbic acid contents of fruits. For instance, Misra et al. (2015) investigated the effect of cold plasma on the chemical quality and ascorbic acid contents of strawberries. In their study, strawberries packaged in polymeric films were exposed to the non-thermal plasma produced from a dielectric barrier discharge. The ascorbic acid content was significantly reduced when the samples were exposed to 80 kV for 1 min. It is interesting to note that the anthocyanin content in the strawberries treated samples was not significantly affected. In another study, the application of cold plasma was reported to affect the degradation of chlorophyll (Ramazzina et al., 2015). A 15% reduction in the chlorophyll content in kiwi fruit was observed after treatment with cold plasma. Chlorophyll degradation by plasma could be mediated by the presence of oxygen produced by cold plasma. In the kiwi fruit, plasma treatment did not affect the ascorbic acid content immediately after plasma treatment, while a significant decrease in the ascorbic acid content observed after storage. Moreover, plasma treatment did not affect the antioxidant content and antioxidant activity of fresh kiwi fruit after cold plasma treatment.

7.2.1.2 Meat Products

Information about the effect of cold plasma on the meat products is very limited. In the meat products, the effects of cold plasma on proteins and lipids is very important. Any changes in the structure of proteins and lipids in meat products directly affect the nutritional and sensorial properties of these products. The free radicals and reactive oxygen species (ROS), including ozone, has the ability to react with fatty acids, oxidizing them to by-products that will affect the nutrient profile of meat products. Plasma has been reported to oxidize lipids by the Criegee mechanism resulting in the production of aldehydes and carboxylic acids in beef (Sarangapani et al., 2017). The extent of lipid oxidation by plasma was reported to be the function of time and applied voltage/energy. Sarangapani et al. (2017) found that the cold plasma treatment of beef significantly decreased the amount of unsaturated fatty acids. This was reported to be due to the cleavage of the double bonds of the unsaturated fatty acids, especially oleic, palmitoleic and linoleic acids by hydroxyl radicals. Oleic acid was reported to be the most susceptible to oxidation by plasma.

In another study, Jayasena et al. (2015) investigated the effect of cold plasma on the quality of pork butt and beef loins. A time-dependent effect of plasma exposure on the extent of lipid oxidation was reported. The lipid oxidation in the pork butt and beef loins was not significantly increased up to a treatment time of 7.5 min. However, a significant increase in the lipid oxidation was observed when the pork butt and beef loins were treated with plasma for 10 min. Kim, Yong, Park, Choe, and Jo (2013) also reported an increase in the lipid oxidation of pork loins treated with plasma produced by a dielectric barrier discharge. Rød, Hansen, Leipold, and Knøchel (2012) observed an increase in the thiobatbituric acid reactive substances (TBARS) of ready-to-eat bresaola treated with cold plasma as a function of plasma power, time and storage.

7.2.1.3 Dairy Products

The effect of plasma on fresh milk was studied by Korachi et al. (2015). In this study, two tungsten electrodes with a 9 kV AC power supply were used. No changes in the fatty acid concentration of the fresh milk compared to the plasma-treated one were observed. After 3 min of plasma treatment, a decrease in the content of unsaturated fatty acids was reported and further decreasing was observed by extending the treatment time. Moreover, slight reductions in stearic acid content were observed after 20 min of plasma treatment. In another study, Korachi et al. (2015) observed an increase in the short-chain length fatty acids of milk after plasma treatment and vice versa for long chain fatty acids. Application of cold plasma up to 20 min did not significantly affect on the fatty acid composition of raw milk. After treating sliced cheddar cheese with flexible thin layer dielectric barrier plasma discharge, Yong et al. (2015), observed a significant increase in TBARS values when the sliced cheddar cheese was exposed to the plasma for 5- and 10-min. The values of 0.132 and 0.183 mg malondialdehyde/kg was reported for untreated and plasma-treated cheddar cheese, respectively.

7.2.2 Textural Properties

7.2.2.1 Fruits and Vegetables

The textural properties of fruits and vegetables are important determinants of their acceptability. Mainly affected by polymeric materials such as pectin and cellulose, in addition to interactions with other components, fruit and vegetable texture can be affected by chemicals, enzymes, storage, and heat. Up to now, only a few studies have reported the effect of cold plasma treatment on the textural properties of fruits and vegetables. Ramazzina et al. (2015) have reported that the cold plasma treatment of kiwi fruit slices did not significantly affect the hardness and the energy needed to rupture the fruit. The penetration tests for the kiwi fruit slices were done at two different points of the pericarp using the texture analyzer.

7.2.2.2 Meat Products

Exposure of meat products to cold plasma does not seem to significantly affect the textural profile. For instance, Jayasena et al. (2015) did not observe any changes in the cohesiveness and springiness of pork butts and beef loins after plasma treatment. Similarly, the hardness, chewiness, and gumminess of the samples did not change, though these parameters have responded more to plasma exposure than springiness and cohesiveness. Kim et al. (2013) also reported that the texture of the pork loins did not significantly change after plasma treatment generated from a flexible thin layer dielectric barrier discharges. More studies are needed to accurately confirm the effects of plasma on the texture of meat products.

7.2.2.3 Dairy Products

There is very little information available on the textural effects of cold plasma on the dairy products. Most of the reports on the use of cold plasma on dairy products have been focused on the decontamination. Thus, more studies need to be conducted to evaluate the effect of cold plasma treatment on the textural properties of dairy products and its feasibility to modify these products.

7.2.3 Sensorial Properties

7.2.3.1 Fruits and Vegetables

Niemira and Sites (2008) did not observe any discoloration, blistering, pitting, unpleasant aroma or any sensory damage after treating golden delicious apples with cold plasma up to 2 min and 40 L/min using a gliding arc cold plasma system. This was reported to be due to the short exposure time of the apples to the plasma discharge. Misra et al. (2014) also found that the color of plasma-treated strawberries was not significantly affected when treated with atmospheric cold plasma. Though reductions were observed in the L, a$*$ and b$*$ values of the plasma-treated strawberries *versus* the untreated samples, the difference was not statistically different. In another study, Shi et al. (2011) observed that the turbidity and pH of freshly squeezed orange juice treated with low-temperature plasma were not significantly affected. Vleugels et al. (2005) measured the effects of atmospheric cold plasma on the color of cold peppers. Red green and yellow peppers were cut into 1-cm portions and exposed to cold plasma up to 20 min. After measuring the extent of discoloration of the samples with a reflectance spectrophotometer, they reported no significant changes in the color of the peppers. Their chroma-hue plots suggested a larger color difference in the green peppers than the other samples.

7.2.3.2 Meat Products

Application of cold plasma has reported affecting the color of meat products. The effect of cold plasma on the color of meat products seems to be related to the time and energy of exposure. Jayasena et al. (2015) reported significant changes in the color parameters (L a∗ and b∗) of pork butts and beef loins after treatment with cold plasma using flexible thin layer dielectric barrier discharge system. The effect of cold plasma on the color was depended on the type of meat product, energy and time of exposure. While the b∗-values of the plasma-treated pork butt was not significantly different from the untreated sample, the b∗-value of the plasma-treated beef loin was significantly higher than the untreated sample. It was also observed that prolonged exposure times led to a decrease in the redness of the pork and an increase in greenness. This decrease in redness of the meat samples could negatively affect the consumer acceptance of plasma-treated meat products, as consumers generally use meat surface color as an indicator of freshness. The increase in the greenness of the meat samples may be due to the reaction of myoglobin with hydrogen peroxide forming choleglobin. A similar observation was made by Fröhling, Baier, Ehlbeck, Knorr, and Schlüter (2012) when fresh pork was treated with cold plasma. The higher b∗-values observed when fresh meat products were treated with cold plasma might be due to the oxidation of deoxymyoglobin or myoglobin, which leads to the formation of metmyoglobin (Jayasena et al., 2015; Mancini & Hunt, 2005). Several studies have reported that the plasma treated fresh pork loins were significantly darker than the non-treated samples (Cheng et al., 2010; Kim et al., 2011; 2013). They reported no difference in the redness and greenness of the meat samples. A reduction in the lightness of the pork loin was attributed to the drying of the surface of the sample due to moisture loss. Ulbin-Figlewicz, Brychcy, and Jarmoluk (2015) and Dirks et al. (2012) did not observe any changes in the color of chicken breast samples treated with cold plasma. The chicken breast samples were treated with plasma up to 3 min. Kim et al. (2013) did not find any significant difference in the consumer acceptability of plasma treated fresh pork sample compared to control. However, Jayasena et al. (2015) reported a negative effect of plasma treatment on the cooked pork butt and beef loins taste. The color, overall acceptability, appearance, and off-odor were not significantly affected. This unacceptable taste perception was observed when cooked pork butts and beef loins were treated for at least 10 min. The negative sensory properties of the cold plasma treated samples could be due to the oxidation of lipids. Lee et al. (2012) reported that the plasma treatment negatively influenced the flavor, taste and overall acceptability of cooked egg yolk but not the egg whites.

7.2.3.3 Dairy Products

Several studies have reported the application of cold plasma treatment on the sensorial properties of dairy products. For instance, Gurol, Ekinci, Aslan, and Korachi (2012) and Kim et al. (2015) did not observe any significant changes in the L and b*- values of commercial whole milk treated with plasma. Even though they observed a decrease in the a*-values of the samples, the decreases were not significant. Cheddar cheese treated with cold plasma resulted in a decrease in the lightness and redness of the samples but was unaffected with respect to the sensory appearance, color, and total color difference. Cold plasma treatment of milk was reported to enhance the lipid oxidation leading to the generation of off-flavors as observed in cheddar cheese (Lee et al., 2012). Yong et al. (2015) reported significant negative effects on the flavor and overall acceptability of sliced cheddar cheese when exposed to the flexible thin-layer dielectric barrier plasma discharge. The off-flavors in the cheese slices could be due to the oxidation of the high fats available in the product, resulting in the production of a variety of lipid oxidation products.

7.3 Conclusions

The effect of cold plasma on the nutritional, texture and sensory characteristics of food products depends on the time and energy of exposure. In the case of fruits and vegetables, cold plasma has significant effects on their bioactive components, especially on the vitamins and anthocyanins. Lipids and proteins are the main nutritional components affected by cold plasma in meat products. The plasma effects on lipids in meat and dairy products lead to the development of off-flavors, affecting the acceptability of these products. The texture was not affected when plasma was applied to fruits, vegetables, and dairy products. It is important to mention that very limited information currently exists in the use of cold plasma for the modification of the nutritional, texture and sensory characteristics of food and hence these conclusions may change in the near future when more information becomes available.

References

Akocak, P. B. (2016). Current progress in advanced research into fungal and mycotoxin inactivation by cold plasma sterilization. In H. Shintani & A. Sakudo (Eds.), *Gas plasma sterilization in microbiology: Theory, applications, pitfalls and new perspectives* (pp. 59–74). Norfolk, UK: Caister Academic Press.

Amini, M., & Ghoranneviss, M. (2016). Effects of cold plasma treatment on antioxidants activity, phenolic contents and shelf life of fresh and dried walnut (*Juglans regia* L.) cultivars during storage. *LWT, 73*, 178–184.

Bahrami, N., Bayliss, D., Chope, G., Penson, S., Perehinec, T., & Fisk, I. D. (2016). Cold plasma: A new technology to modify wheat flour functionality. *Food Chemistry, 202*, 247–253.

Bastos, D. C., dos Santos, A. E. F., & Simao, R. A. (2014). Acetylene coating on cornstarch plastics produced by cold plasma technology. *Starch, 66*(3–4), 267–273.

Brandenburg, R., Ehlbeck, J., Stieber, M., Woedtke, T. v., Zeymer, J., Schlüter, O., & Weltmann, K.-D. (2007). Antimicrobial treatment of heat sensitive materials by means of atmospheric pressure Rf-driven plasma jet. *Contributions to Plasma Physics, 47*(1-2), 72–79.

Bursać Kovačević, D., Putnik, P., Dragović-Uzelac, V., Pedisić, S., Režek Jambrak, A., & Herceg, Z. (2016). Effects of cold atmospheric gas phase plasma on anthocyanins and color in pomegranate juice. *Food Chemistry, 190*, 317–323.

Cheng, S. Y., Yuen, C. W. M., Kan, C. W., Cheuk, K. K. L., Daoud, W. A., Lam, P. L., & Tsoi, W. Y. I. (2010). Influence of atmospheric pressure plasma treatment on various fibrous materials: Performance properties and surface adhesion analysis. *Vacuum, 84*(12), 1466–1470.

Dirks, B. P., Dobrynin, D., Fridman, G., Mukhin, Y., Fridman, A., & Quinlan, J. J. (2012). Treatment of raw poultry with nonthermal dielectric barrier discharge plasma to reduce *Campylobacter jejuni* and *Salmonella enterica. Journal of Food Protection, 75*(1), 22–28.

Dong, S., Gao, A., Xu, H., & Chen, Y. (2017). Effects of dielectric barrier discharges (DBD) cold plasma treatment on physicochemical and structural properties of zein powders. *Food and Bioprocess Technology, 10*(3), 434–444.

Elez Garofulić, I., Režek Jambrak, A., Milošević, S., Dragović-Uzelac, V., Zorić, Z., & Herceg, Z. (2015). The effect of gas phase plasma treatment on the anthocyanin and phenolic acid content of sour cherry Marasca (*Prunus cerasus* var. Marasca) juice. *LWT – Food Science and Technology, 62*(1, Part 2), 894–900.

Fröhling, A., Baier, M., Ehlbeck, J., Knorr, D., & Schlüter, O. (2012). Atmospheric pressure plasma treatment of *Listeria innocua* and *Escherichia coli* at polysaccharide surfaces: Inactivation kinetics and flow cytometric characterization. *Innovative Food Science & Emerging Technologies, 13*, 142–150.

Grzegorzewski, F., Ehlbeck, J., Schlüter, O., Kroh, L. W., & Rohn, S. (2011). Treating lamb's lettuce with a cold plasma – Influence of atmospheric pressure Ar plasma immanent species on the phenolic profile of *Valerianella locusta. LWT – Food Science and Technology, 44*(10), 2285–2289.

Gurol, C., Ekinci, F. Y., Aslan, N., & Korachi, M. (2012). Low temperature plasma for decontamination of *E. coli* in milk. *International Journal of Food Microbiology, 157*(1), 1–5. https://doi.org/10.1016/j.ijfoodmicro.2012.02.016

Han, L., Boehm, D., Amias, E., Milosavljević, V., Cullen, P. J., & Bourke, P. (2016). Atmospheric cold plasma interactions with modified atmosphere packaging inducer gases for safe food preservation. *Innovative Food Science & Emerging Technologies, 38*, 384–392.

Jayasena, D. D., Kim, H. J., Yong, H. I., Park, S., Kim, K., Choe, W., & Jo, C. (2015). Flexible thin-layer dielectric barrier discharge plasma treatment of pork butt and beef loin: Effects on pathogen inactivation and meat-quality attributes. *Food Microbiology, 46*, 51–57.

Kim, B., Yun, H., Jung, S., Jung, Y., Jung, H., Choe, W., & Jo, C. (2011). Effect of atmospheric pressure plasma on inactivation of pathogens inoculated onto bacon using two different gas compositions. *Food Microbiology, 28*(1), 9–13.

Kim, H.-J., Yong, H. I., Park, S., Choe, W., & Jo, C. (2013). Effects of dielectric barrier discharge plasma on pathogen inactivation and the physicochemical and sensory characteristics of pork loin. *Current Applied Physics, 13*(7), 1420–1425.

Kim, H.-J., Yong, H. I., Park, S., Kim, K., Choe, W., & Jo, C. (2015). Microbial safety and quality attributes of milk following treatment with atmospheric pressure encapsulated dielectric barrier discharge plasma. *Food Control, 47*, 451–456.

Kim, H.-S., & Min, S. C. (2017). Effects of microwave-discharged cold plasma on synthesis and characteristics of citrate derivatives of corn starch granules. *Food Science and Biotechnology, 26*(3), 697–706.

Korachi, M., Ozen, F., Aslan, N., Vannini, L., Guerzoni, M. E., Gottardi, D., & Ekinci, F. Y. (2015). Biochemical changes to milk following treatment by a novel, cold atmospheric plasma system. *International Dairy Journal, 42*, 64–69.

Lee, H.-J., Jung, S., Jung, H.-S., Park, S.-H., Choe, W.-H., Ham, J.-S., & Jo, C. (2012). Evaluation of a dielectric barrier discharge plasma system for inactivating pathogens on cheese slices. *Journal of Animal Science and Technology, 54*(3), 191–198.

Los, A., Ziuzina, D., Boehm, D., Cullen, P. J., & Bourke, P. (2017). The potential of atmospheric air cold plasma for control of bacterial contaminants relevant to cereal grain production. *Innovative Food Science & Emerging Technologies, 44*, 36–45.

Mancini, R. A., & Hunt, M. C. (2005). Current research in meat color. *Meat Science, 71*(1), 100–121.
Misra, N. N., & Jo, C. (2017). Applications of cold plasma technology for microbiological safety in meat industry. *Trends in Food Science & Technology, 64*, 74–86.
Misra, N. N., Kaur, S., Tiwari, B. K., Kaur, A., Singh, N., & Cullen, P. J. (2015). Atmospheric pressure cold plasma (ACP) treatment of wheat flour. *Food Hydrocolloids, 44*, 115–121.
Misra, N. N., Patil, S., Moiseev, T., Bourke, P., Mosnier, J. P., Keener, K. M., & Cullen, P. J. (2014). In-package atmospheric pressure cold plasma treatment of strawberries. *Journal of Food Engineering, 125*, 131–138.
Niemira, B. A., & Sites, J. (2008). Cold plasma inactivates *Salmonella Stanley* and *Escherichia coli* O157: H7 inoculated on golden delicious apples. *Journal of Food Protection, 71*(7), 1357–1365.
Ramazzina, I., Berardinelli, A., Rizzi, F., Tappi, S., Ragni, L., Sacchetti, G., & Rocculi, P. (2015). Effect of cold plasma treatment on physico-chemical parameters and antioxidant activity of minimally processed kiwifruit. *Postharvest Biology and Technology, 107*, 55–65.
Rød, S. K., Hansen, F., Leipold, F., & Knøchel, S. (2012). Cold atmospheric pressure plasma treatment of ready-to-eat meat: Inactivation of *Listeria innocua* and changes in product quality. *Food Microbiology, 30*(1), 233–238.
Sarangapani, C., Ryan Keogh, D., Dunne, J., Bourke, P., & Cullen, P. J. (2017). Characterisation of cold plasma treated beef and dairy lipids using spectroscopic and chromatographic methods. *Food Chemistry, 235*, 324–333.
Scholtz, V., Pazlarova, J., Souskova, H., Khun, J., & Julak, J. (2015). Nonthermal plasma — A tool for decontamination and disinfection. *Biotechnology Advances, 33*(6, Part 2), 1108–1119.
Selcuk, M., Oksuz, L., & Basaran, P. (2008). Decontamination of grains and legumes infected with *Aspergillus* spp. and *Penicillum* spp. by cold plasma treatment. *Bioresource Technology, 99*(11), 5104–5109.
Shi, H., Cooper, B., Stroshine, R. L., Ileleji, K. E., & Keener, K. M. (2017). Structures of degradation products and degradation pathways of aflatoxin B-1 by high-voltage atmospheric cold plasma (HVACP) treatment. *Journal of Agricultural and Food Chemistry, 65*(30), 6222–6230.
Shi, X., Zhang, G., Wu, X., Li, Y., Ma, Y., & Shao, X. (2011). Effect of low-temperature plasma on microorganism inactivation and quality of freshly squeezed orange juice. *IEEE Transactions on Plasma Science, 39*(7), 1591–1597.
Takai, E., Kitamura, T., Kuwabara, J., Ikawa, S., Yoshizawa, S., Shiraki, K., … Kitano, K. (2014). Chemical modification of amino acids by atmospheric-pressure cold plasma in aqueous solution. *Journal of Physics D: Applied Physics, 47*(28), 285403.
Thirumdas, R., Trimukhe, A., Deshmukh, R. R., & Annapure, U. S. (2017). Functional and rheological properties of cold plasma treated rice starch. *Carbohydrate Polymers, 157*, 1723–1731.
Ulbin-Figlewicz, N., Brychcy, E., & Jarmoluk, A. (2015). Effect of low-pressure cold plasma on surface microflora of meat and quality attributes. *Journal of Food Science and Technology, 52*(2), 1228–1232.
Vleugels, M., Shama, G., Deng, X. T., Greenacre, E., Brocklehurst, T., & Kong, M. G. (2005). Atmospheric plasma inactivation of biofilm-forming bacteria for food safety control. *IEEE Transactions on Plasma Science, 33*(2), 824–828.
Yasuda, H., Miura, T., Kurita, H., Takashima, K., & Mizuno, A. (2010). Biological evaluation of DNA damage in bacteriophages inactivated by atmospheric pressure cold plasma. *Plasma Processes and Polymers, 7*(3–4), 301–308.
Yong, H. I., Kim, H.-J., Park, S., Kim, K., Choe, W., Yoo, S. J., & Jo, C. (2015). Pathogen inactivation and quality changes in sliced cheddar cheese treated using flexible thin-layer dielectric barrier discharge plasma. *Food Research International, 69*, 57–63.

Chapter 8
Impact of Ultraviolet Processing on Food Composition

María Lavilla, Amaia Lasagabaster, and Iñigo Martínez-de-Marañón

8.1 Introduction

Ultraviolet (UV) radiation comprises the wavelength from 100 to 400 nm in the spectrum of electromagnetic waves. Likewise, the ultraviolet spectrum can be distinguished between UV-A (400-315 nm), UV-B (315-280 nm), and UV-C (280-100 nm) (Fig. 8.1). Among the complete UV range, the UV-C radiation, and more specifically, the wavelength at 254 nm, has been demonstrated to achieve the highest germicidal effect (Ma, Zhang, Bhandari, & Gao, 2017). Indeed, UV-C radiation has been confirmed to be a useful tool to inactivate bacteria (including spores), viruses, yeast, molds, and parasites (Gayán, Condón, & Álvarez, 2014; Gómez-López, Koutchma, & Linden, 2012). This high effectivity is due to the absorption of this radiation by DNA/RNA (Gayán et al., 2014) and proteins (Díaz, Candia, & Cobos, 2017; Kuan, Bhat, Patras, & Karim, 2013; Pellicer & Gómez-López, 2017), which provokes a loss of structure and subsequent malfunction. In this chapter, the impact of ultraviolet processing on food constituents was reviewed.

However, the efficiency of the microbial inactivation by UV radiation is influenced by many parameters, principally related to the transmitted energy form and to the specifications of the treated product. More specifically, regarding the transmitted energy form, the crucial parameters are the power, the wavelength, the mode of irradiation (continuous or pulsed), and the treatment time. Regarding the treated matrix, the most important factors are the state (solid or fluid), the dimensions (shape and thickness), the physical properties (e.g., density and viscosity, transparency to UV-light, etc.), and the product's composition (Fan, Huang, & Chen, 2017; Koutchma, 2008, 2009). Moreover, microbial species and their characteristics, are also actors to be considered, since the UV-C dosage extremely differs for inhibiting

M. Lavilla (✉) · A. Lasagabaster · I. Martínez-de-Marañón
AZTI, Department of Food Quality, Safety and Identity, Technological Park Bizkaia, Derio (Bizkaia), Spain
e-mail: mlavilla@azti.es

S. Roohinejad et al. (eds.), *Effect of Emerging Processing Methods on the Food Quality*,
https://doi.org/10.1007/978-3-030-18191-8_8

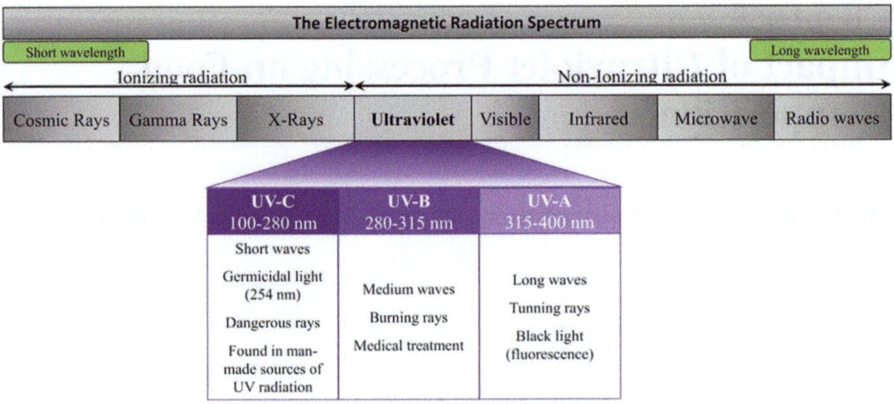

Fig. 8.1 The electromagnetic radiation spectrum subdivision and characteristics of UV radiation

different microorganisms (Guerrero-Beltrán & Barbosa-Cánovas, 2004; Koutchma, 2009). Consequently, although UV-C treatment has the potential to eliminate undesirable microorganisms in foods, all these influencing parameters make it difficult to achieve full control of the variables associated with process operations, bringing limitations on the applications of UV for further disinfection processes.

In the food industry, UV-C has been usually used for the decontamination of water supplies to obtain water with high-microbiological quality as raw material and for cleaning air particle filters, equipment, and other food contact surfaces (Ohlsson, 2002). In fact, the use of UV-C disinfection in drinking and wastewater as an alternative to chlorination has experienced a great increase over the last few years. Indeed, this technology is currently very common in drinking water supplies in North America and Europe (Pereira & Vicente, 2010), and it is expected to incessant growth with the development of more efficient units. Apart from that, UV-C light is also one of the technologies frequently used for sterilization of food packages under aseptic conditions (Ozen & Floros, 2001).

Moreover, UV-C radiation can be considered as a promising non-thermal technology for further applications in high-quality food preservation (Guerrero-Beltrán & Barbosa-Cánovas, 2004). This technology was approved by the US Food and Drug Administration for the processing of juices and solid foods (USFDA, 2000). The European Food Safety Authority (EFSA) has recently given the green light to treat milk and bread by UV-C radiation, in order to boost vitamin D and extend shelf-life, respectively (EFSA, 2015, 2016). For further food applications in Europe, however, the European novel food regulation (1997) must be followed (European Commission, 1997). This legislation specifies that "a food should be considered a novel food where it results from a production process not used for food production within the Union before 15 May 1997, which gives rise to significant changes in the composition or structure of food affecting its nutritional value, metabolism or level of undesirable substances". Consequently, each specific food or food ingredients treated with UV light would be reviewed to decide if it falls under the remit of the novel food legislation.

In conclusion, UV technology shows promising advantages when applied to food processing for reducing microbial risks, while theoretically minimizing the loss of quality in terms of flavor, color, and nutritional value compared to thermal treatment. In this chapter, the main objective is to review and summarize the use of UV light in foods, focus on its advantages and effects in food quality, and retention of nutrients and sensory attributes. In addition, the potential negative effects, limitations and outstanding challenges for commercial application of this innovative technology, including valuable opportunities for increasing its implementation in the food industry, are comprised.

8.2 Advantageous and Challenge of Using Ultraviolet on Food Quality

UV-C light has been widely studied in terms of microbial inactivation efficiency. Taking into account the aforementioned parameters influencing effectivity and the low penetration depth of the UV light, it has been traditionally limited to the applications to fluid food surfaces: liquids with high concentration of color compounds, organic solutes and suspended matter tend to absorb and scatter UV photons limiting the penetrating power of the radiation in the surface (Koutchma, 2008, 2009). In the case of solid foods, an irregular surface may form shadow areas in the product that may protect microorganisms from the radiation (Manzocco & Nicoli, 2015).

Nevertheless, thanks to some research, UV-based technology has successfully demonstrated its potential as a feasible and safe processing method for a wider variety of foods. Moreover, it has been checked that most of the physicochemical properties and sensory attributes of food products remain almost without changes after UV processing at reasonable doses (Guerrero-Beltrán & Barbosa-Cánovas, 2004). Accordingly, the application of UV-C processing has been tested in liquid foods, such as milk, liquid egg, wine, beer, and animal blood. Other factual applications are the treatment of sugar syrups for inactivation of dormant spores and to recycle brines saving production costs (Kershner, 2015). For solids, UV-C systems are already available to decontaminate eggs and eggshells, fresh and frozen vegetables, meat, hamburger patties, seafood, bread, refrigerated pasta, cheese, and other ready-to-eat products (EFSA, 2005, 2015; Koutchma, Forney, & Moraru, 2009).

With these data, it looks that the traditional restriction of the use of UV-C technology to surface treatment and clear and low-turbid foods can be overcome. Consequently, the decontamination of turbid products has been also checked, mostly using the combination of UV-C light with other technologies (García Carrillo, Ferrario, & Guerrero, 2017; Gouma, Álvarez, Condón, & Gayán, 2015). In addition, powdered and granular products (e.g., powdered infant formula or cacao powder) have been proposed by UV-C, despite the encountered difficulties (Arroyo et al., 2017; Stoops, Jansen, Claes, & Van Campenhout, 2013).

Furthermore, consumers demand emerging technologies to tend to the development of new shelf-stable products with minimal product quality deterioration. Although UV-C light might inactivate bacterial spores, a higher dosage is needed to reach an acceptable inactivation compared to vegetative cells. Consequently, and together with the low UV light transmission in food products, studies about the combination of UV-C technology with other strategies (*hurdle technology*) in order to reduce the total treatment intensity and product damage, as well as to improve microbial control, are also increasing. The use of UV-C combined with soft thermal processing (García Carrillo et al., 2017; Marquenie, Michiels, Van Impe, Schrevens, & Nicolaï, 2003), hydrogen peroxide spraying (Birmpa, Sfika, & Vantarakis, 2013), organic acids (Chen, Hu, He, Jiang, & Zhang, 2016; Nogales-Delgado, Fernández-León, Delgado-Adámez, Hernández-Méndez, & Bohoyo-Gil, 2014; Raybaudi-Massilia, Calderón-Gabaldón, Mosqueda-Melgar, & Tapia, 2013), modified atmosphere packaging (Choi, Yoo, & Kang, 2015), gaseous ozone (Gutiérrez, Chaves, & Rodríguez, 2017) and electrolyzed water (Martínez-Hernández et al., 2015; Santo, Graça, Nunes, & Quintas, 2016), among others, have been studied. In these studies, an improvement in microbial inactivation due to the synergy between UV and the other applied antimicrobials has been confirmed.

As seen, there is increasing knowledge about UV-light effects in bacterial inactivation. However, there are still many research needs for ultraviolet radiation, such as more comprehensive information about inactivation kinetics for pathogens and spoilage microorganisms, develop new methods to measure dose-response behavior of microorganisms in viscous and absorptive products and develop validation methods (Koutchma, 2008; Koutchma & Parisi, 2004).

Nevertheless, the applications of UV light in food go beyond microbial decontamination. This technology can also be used to degrade chemical contaminants in foods without residues, such as aflatoxins, mycotoxins, dioxins and some pesticides by direct photolysis due to their potential to absorb light (Baron, Børresen, & Jacobsen, 2005; Diao et al., 2015; Misra, 2015; Tripathi & Mishra, 2010; Zhu, Koutchma, Warriner, & Zhou, 2014). Furthermore, UV-C can effectively inactivate spoilage enzymes of different food matrices (Aguilar, Ibarz, Garvín, & Ibarz, 2016; Augusto, Ibarz, Garvín, & Ibarz, 2015). However, as in the case of microbial inactivation, the effectiveness of UV light to reduce chemicals and enzymatic spoilage are lower in turbid environments than in clear solutions, and these could be one of the main reasons for the limited application of UV in the food field. Consequently, further research about this technology is still necessary to guarantee an accurate uniform treatment in all the locations of the product.

Apart from these cited advantages and challenges, the environmental footprint of technology is also an important issue for the food industry. From a technological point of view, continuous UV-C radiation is distinctive for a lower cost of equipment and lower energy consumption in comparison to high-temperature short-time (HTST), PEF, and HHP treatments (Basaran, Quintero-Ramos, Moake, Churey, & Worobo, 2004; Rodriguez-Gonzalez, Buckow, Koutchma, & Balasubramaniam, 2015). Furthermore, as far as it is known, it neither generates toxic by-products or effects nor detrimental residues for the environment (Guerrero-Beltrán & Barbosa-Cánovas, 2004). Another advantage of this procedure is the relatively inexpensive and easy-to-use equipment needed (Bintsis, Litopoulou-Tzanetaki, & Robinson, 2000). In this line, UV-light emitting diodes (LEDs) have been developed increasing advantages such as low cost, energy-efficient and long-life expectation. Also, contrarily to low-pressure mercury vapor lamps (emitting continuous UV radiation at 185 and 254 nm) currently allowed for juice processing (USFDA, 2001), LEDs cause no harm for human eyes and skin and do not create mercury waste (Mori et al., 2007). Finally, as a preservation method, UV-based technologies have a positive consumer image because of their multiple cited advantages (Lavilla & Gayán, 2018).

Consequently, food specialists consider UV light/radiation processing on the top innovative technologies with most potential, and anticipate an increase of commercial application of UV treatment in the next 5–10 years, especially in the drink and beverage and fresh produce sectors (Jermann, Koutchma, Margas, Leadley, & Ros-Polski, 2015), together with other technologies as high-pressure processing (HPP), microwave (MW) pasteurization/sterilization, and Pulsed Electric Fields (PEFs).

In conclusion, there is an increasing interest in using UV-C light in food industry applications. However, besides its desirable antimicrobial and technical properties, the impact of this technology in food characteristics must be carefully revised, regarding the high variability in results depending on the food type and UV dosage. The influence of UV-light in nutritional, textural and sensory characteristics of fruits, vegetables, meat, poultry, fish and dairy products are addressed in the following sections.

8.2.1 Nutritional Properties

As previously reported, in general, processing of foods with UV light technology has demonstrated its potential to preserve most of the physicochemical attributes of products (Caminiti et al., 2012; Taze, Unluturk, Buzrul, & Alpas, 2015). However, UV irradiation is absorbed by UV-sensitive food components, and this absorption initiates chemical reactions by inducing changes in the molecular bonds or by photosensitizing the molecules (Koutchma et al., 2009). Therefore, there are obvious changes in the chemical (and nutritional) composition of food components and product quality deterioration after a UV-light treatment, especially if it is applied in high doses. The most sensitive molecules are nucleic acids, proteins with aromatic amino acids (phenylalanine, tryptophan and tyrosine), proteins with disulfide bonds (cystine) and nutrients such as vitamins (A, B_2, B_9, B_{12}, C, D, E, K) and unsaturated fatty acids and phospholipids (Koutchma, 2009; Spikes, 1981). A brief summary of this potential, and sometimes contradictory effects of UV light in nutritional components of foods are listed in Table 8.1. However, the photosensitivity of molecules highly varies depending on the molecule itself and the wavelength they absorb but also depending on food composition (e.g. presence of some food pigments or other photoreactive groups) (Koutchma et al., 2009). Thus, each food group is considered below, evaluating UV light effects in their nutritional characteristics.

Table 8.1 Potential positive and negative effects of UV-light on the nutritional quality of foods

Positive effects	Undesirable effects
Increase synthesis of vitamin D Increase antioxidant capacity and bioactive compounds concentration Destruction of allergens and reduction of IgE immunoreactivity Increase extractability of bioactive compounds Destruction of chemical contaminants (mycotoxins, dioxins…) Increase protein digestibility Increase of antioxidant and antihypertensive activities of the α-casein	Destruction of light-sensitive vitamins. Possible production of furan and free radicals in foods rich in fructose (application requires chemical and toxicological evaluation) Increase allergenic potential Destruction of unsaturated fatty acids (peroxidation) Increased acid values and peroxide values Cross linking in carbohydrate and protein Potential formation of furan and/or hydroxyalkyl and acyl radicals Increase of biogenic amines Formation of radiolytic products in lipid –containing foods

8.2.1.1 Fruits and Vegetables

Fruit juice processing is one of the most common applications of UV technologies in the food industry. This technology offers a safe product with extended shelf life and, compared to thermally pasteurized juice, a superior organoleptic and nutritional quality. UV-C radiation is used for the treatment of apple juice and cider since UV-processing units are more affordable than heat pasteurizers (Basaran et al., 2004). In the field of solid foods, UV-C is particularly suitable for surface decontamination of fruits and vegetables in both whole and fresh-cut formats (Fan et al., 2017; Fonseca & Rushing, 2006).

Apart from microbial inactivation and subsequent prolonged shelf life, nutritional changes may be obtained from the application of UV treatment in fruits and vegetables since, due to its nature, fruits and vegetables contain a high content in health-promoting but light-sensitive compounds (Koutchma, Popović, Ros-Polski, & Popielarz, 2016).

By one hand, nutritional benefits have been reported by several works as the increase of antioxidant capacity and the better extractability of bioactive compounds (e.g., carotenoids, lycopene, and phenols) in juice and plant tissues (Bravo et al., 2013; Santhirasegaram, Razali, George, & Somasundram, 2015).

More specifically, exposure to UV induces the synthesis of health-promoting compounds such as resveratrol in grapes (Cantos, Espín, & Tomás-Barberán, 2001), and total soluble phenolics in sliced parsnip and fresh-cut lettuce (Du, Avena-Bustillos, Breksa, & McHugh, 2014). Similarly, UV also contributed to significant increases in total soluble phenolics, total phenolics, and total antioxidant capacity of fresh-cut carrot (Du, Avena-Bustillos, Breksa, & McHugh, 2012; Formica-Oliveira, Martínez-Hernández, Díaz-López, Artés, & Artés-Hernández, 2017) and tomatoes (Vunnam et al., 2014). Furthermore, UV-C was also successful in increasing antioxidants (total flavonoid, reducing power and ABTS scavenging activity) in mangoes and pineapples (George, Razali, Santhirasegaram, & Somasundram, 2015,

2016), and total anthocyanin, phenolic compounds, L-ascorbic acid content, and volatile compounds in strawberries (Severo, de Oliveira, Tiecher, Chaves, & Rombaldi, 2015). On the other hand, UV-A-LED has been demonstrated to contribute to the increase in quercetin glycoside content of watercress (Kanazawa, Hashimoto, Yoshida, Sungwon, & Fukuda, 2012), and in anthocyanin, vitamin C, and total phenolics of strawberry (Kim, Bae, & Chun, 2011). In conclusion, UV-C treatment is able to induce a variety of positive changes in the nutritional composition of some fruits, resulting in higher levels of bioactive molecules. In other cases, maintenance of beneficial nutrients has been reported when fruit and vegetable products are treated by UV-C, which reflects one of the potential benefits of this technology compared to thermal treatments: UV radiation does not significantly degrade neither polyphenols in apple juice (Islam et al., 2016), nor total phenolic compounds and vitamin C contents in fresh-cut paprika (Choi et al., 2015), and mandarin (Shen et al., 2013).

Regarding the extractability of these valuable compounds, authors have also found an increase in nutrients: for instance, *trans*-resveratrol, *trans*-piceid, *p*-coumaric-, caffeic-, ferulic- acids, and total phenolic content in peanuts achieved their greatest extractability when a combination of ultrasounds and UV radiation treatment was applied (Sales & Resurreccion, 2010). The use of continuous UV-C light also favored the extraction of anthocyanins, tannins, and aromatic compounds from fox grapes (Fava et al., 2011), although in this case, extraction may be favored by micro-damage in skin cells by UV-C irradiation.

Further "nutritional" benefits of UV-light technologies include the decrease of IgE-binding by several important fruit/vegetable allergens, such as in peanuts (Chung, Yang, & Krishnamurthy, 2008; Zhao et al., 2014), soybean (Yang et al., 2010) and mango (George et al., 2016), which opens the opportunity to develop hypoallergenic fruit products.

All of these results demonstrate the effects of UV-C treatment for maintaining or increasing the nutritional quality of certain fruits as well as the potential of this treatment in shelf life extension. However, in contrast to these benefits, other works have concluded that photosensitive components could be destroyed. For instance, in orange juice vitamins B_1 (thiamine), B_2 (riboflavin) and β-carotene are partially destroyed by UV treatment (Koutchma et al., 2009). Vitamin C is also a light-sensitive vitamin in fruit and vegetable juices and can be degraded by UV irradiation (Koutchma, 2008; Koutchma, Keller, Chirtel, & Parisi, 2004; Pan & Zu, 2012). Although vitamin retention depends on the composition of the food matrix, UV source, and dose, similar retention of ascorbic acid compared to that of thermal pasteurization can be obtained, and consequently, regarding only this nutritional parameter, UV technologies do not provide an advantage (Orlowska et al., 2012; Santhirasegaram et al., 2015). Similarly, UV-radiation induced significant losses of ascorbic acid and carotene content in red chili powder (Tripathi & Mishra, 2010) and small losses of anthocyanins in grapes (Pala & Toklucu, 2013), but less than those observed by heat treatment. In apple juice, although polyphenols seem to be preserved, significant decreases in several important vitamin concentrations were observed (Islam et al., 2016).

Apart from vitamins, other nutrients such as oleuropein, an antioxidant from olives, may loss its stability under UV-C light, suffering a series of fast decomposition reactions leading to hydroxytyrosol (one of the main phenolic components of olive oil) and elenolic acid (a marker for maturation of olives) (Longo, Morozova, & Scampicchio, 2017). Also, in other vegetable oils, UV irradiation can slightly increase the acid and peroxide values of the treated samples, and also destroy the unsaturated fatty acids in various degrees (Shen et al., 2014). Apart from nutrient degradation, in fruits, high UV-C doses also cause other negative nutritional disadvantages, as they may induce fructose photolysis which can lead to the formation of furan and/or hydroxyalkyl and acyl radicals (Orlowska et al., 2012). Also, UV-C treatment seems to induce accumulation of the allergenic protein Fra a1 (strawberry) (Severo et al., 2015).

8.2.1.2 Meat Products

UV treatments have been also applied for surface decontamination of meat products, although these applications have not been so broadly considered (Heinrich, Zunabovic, Varzakas, Bergmair, & Kneifel, 2016; Koutchma & Orlowska, 2012). In general, the studies have concluded that UV light can be successfully applied to meats, poultry, and fish products, preserving their quality and nutritional attributes. However, when considered more specifically, UV-C radiation possibly affects the physicochemical properties of meat products, and several undesirable effects may be observed in nutritional characteristics, concerning the expected changes of photosensitive components.

Studies on chicken breasts showed that UV-C radiated products exhibited an increase in tyramine, cadaverine, and putrescine contents (Lázaro et al., 2014). Similar results, as well as an increase in histamine content, have been found in hybrid "cachamay" fish filets (de Oliveira Bottino, Rodrigues, de Nunes Ribeiro, Lázaro, & Conte-Junior, 2016). As UV-light reduces the number of microorganisms in both products, this increase on the biogenic amines content should be considered as an effect of the UV processing and not as an indicator of bacterial growth (Lázaro et al., 2014).

An important and shocking issue is the recent findings of the formation of 2-alkylcyclobutanones (2-ACBs) by UV-light. These chemical compounds are considered as unique radiolytic products in lipid-containing foods that could only be formed through exposure to ionizing radiation and consequently are currently the marker molecules required by the European Committee for Standardization to be used for the identification of foods irradiated with ionizing irradiation. However, the generation of 2-ACBs is also possible when fatty acids and triglycerides are exposed to UV-C light source in corn oil and pork samples (Meng & Chan, 2017).

Concerning the lipid stability, results show that alterations are dose-dependent. Park and Ha (2015) found that UV-C radiation at intensities between 1800–3600 mWs/cm^2 can cause an increase in the lipid peroxidation of the chicken breast meat. However, other authors have recently found no significant difference between non-treated and UV-C treated samples up to 2400 mWs/cm^2 (Yang, Sadekuzzaman, & Ha, 2017).

In addition, a low UV-C dose is suggested for tilapia fish fillets as an alternative processing method to control the formation of biogenic amines (Monteiro et al., 2017). According to these authors, although the UV-C treatment increased protein oxidation, lipid oxidation was not influenced, and positive improvement in the total polyunsaturated fatty acid quantity and greater overall nutritional quality was noticed.

In other protein-rich products, such as in egg, vitamins (B_2, B_5, and E) are UV-light stable. However, as occurred in some fruits, retinal, vitamin C and carotenoids are degraded, showing losses up to 80%, 66% and 61%, respectively (de Souza et al., 2015). Also in eggs, ultraviolet light has demonstrated other benefits such as reduction of the potential allergenicity of egg proteins (Anugu, Yang, Shriver, Chung, & Percival, 2010).

8.2.1.3 Dairy Products

UV-based processing of milk has been recently approved by the European Union following European Commission novel food legislation (EFSA, 2016). In this line, UV has largely shown its potential to improve milk's microbiological quality (Datta, Harimurugan, & Palombo, 2015) while no negative or toxigenic effects have been elucidated. Also, in the nutritional field, EFSA experts' dossier highlights that UV radiation has an important benefit, by increasing vitamin D levels in milk. In more recent studies, other nutritional benefits such an increase of antioxidant and antihypertensive activities of α-casein have been also assessed after UV-light treatment (Hu et al., 2017).

Nonetheless, although approved, the application of UV pasteurization in the dairy industry is still a major challenge due to several reasons. By one hand, the high-intensity treatment needed to compensate low milk's UV penetration capacity may have the consequent appearance of sensorial defects, which will be described in the following section of this chapter. On the other hand, contradictory results regarding nutritional aspects are still found in the literature. Contrarily to cited advantages, several studies have shown that vitamins C, E, A, B_2, and even vitamin D are sensitive to UV-light in milk matrices, and their content can decrease due to intense treatments (Cappozzo, Koutchma, & Barnes, 2015; Guneser & Karagul Yuceer, 2012). However, the content of riboflavin (B_2) and vitamin B_{12} contents are similar in UV-treated and heat-pasteurized milk (EFSA, 2016). Moreover, oxidations of proteins and unsaturated lipids have been observed (Rossitto et al., 2012) although these effects may not be statistically significant (Cappozzo et al., 2015). Consequently, in general, this novel food processing has not been considered as nutritionally disadvantageous by the expert panel, when treated at the proper UV-dose (1000–2000 J/L).

Milk from other species, such as goat milk, has also been treated by UV-light with similar results in milk nutritional modifications, such as increased oxidation and hydrolytic rancidity (Guneser & Karagul Yuceer, 2012; Matak et al., 2007).

Additionally, taking into account that cow's milk allergy is one of the most important and most frequent allergy worldwide (Fiocchi et al., 2010), it deserves special attention the fact that several studies have found a significant reduction in IgE binding values compared to control samples, indicating reduction in allergenicity of milk proteins (Hu et al., 2017; Tammineedi, Choudhary, Perez-Alvarado, & Watson, 2013). This decreased allergenicity is the consequence of a modification in the structure of proteins by UV-C light, which also leads to an improved digestibility (Hu et al., 2017). However, all of these results have been demonstrated *in vitro* and additional research studies using *in vivo* clinical trials must be carried out in order to confirm these promising results.

8.2.2 Textural Properties

The application of UV-light to foods may have an impact on all photosensitive components, which also include proteins and other structures of the cells, as mentioned previously. Although UV-light-sensitive proteins represent only 10% of the total proteins in foods (Koutchma et al., 2009), changes and aggregation caused by UV-light may have an important effect on the texture, flavor, and appearance. Although not directly related to food texture, it is worth noting at this point that the well-known UV-induced protein cross-linking leads to another important application of UV-light for industry beyond decontamination, as this technology may be extensively applied on the production of protein films. Several authors have reported the production of improved film structures with a wide-ranging of food-origin proteins: whey proteins, soy, gluten, or albumin protein films, among others, have shown better barrier properties after exposure to UV irradiation (Díaz et al., 2017; Park, Cho, & Rhee, 2003; Rhim, Gennadios, Fu, Weller, & Hanna, 1999).

Furthermore, the impact on the structure of the cell wall by UV-light has been evidenced in several works to also impact the surface texture of treated products. Contrarily, the inhibition of microbial growth and delay of other degradation pathways (e.g., inhibition of enzymes affecting texture) may have an indirect positive effect, by maintaining the texture of the treated products. These potential effects are thoroughly reviewed in the subsequent sections.

8.2.2.1 Fruits and Vegetables

In the case of fruits and vegetables, the maintenance of firmness is highly desired because a crunchy texture is commonly associated with freshness. Reported results about the effect of UV light in the texture of fruits and vegetables are very variable, although in most cases, UV-light technology shows to have neither negative nor positive direct effect in this characteristic of fruits, showing the advantage of this technology over thermal treatment. However, UV tends to positively influence the texture of fruits and vegetables indirectly. In these cases, the application of ultraviolet light delayed the softening of the treated fruits, resulting in a firmer texture for a longer period compared to the untreated fruits. For instance, in strawberries, UV applied at high doses had a positive influence on both fruit and flesh firmness (Marquenie et al., 2002; Severo et al., 2015). This effect has been also evidenced in other fruits such as melon and mango, where instrumental texture measurements confirmed a better firmness retention in fruit treated under UV-C radiation, leading to an extended shelf life when compared to control (untreated) samples (Lamikanra, Kueneman, Ukuku, & Bett-Garber, 2005; Promyou & Supapvanich, 2016). Regarding vegetables, UV-treated lettuce and peppers are not adversely affected by the treatment and maintain their texture quality during the studied storage period at refrigeration temperatures (Kim et al., 2013; Nogales-Delgado et al., 2014; Rodoni, Zaro, Hasperué, Concellón, & Vicente, 2015).

Additionally, UV has shown to be a viable alternative in highly perishable fresh shiitake mushrooms, resulting in a reduced decrease in firmness during shelf life storage, and even maintenance of a high level of firmness during 15 days at low temperatures (Jiang, Jahangir, Jiang, Lu, & Ying, 2010). Tomato fruit has been systematically studied in numerous studies, confirming again a better resistance to decay and a slower ripening upon light exposure (Liu et al., 1993; Mukhopadhyay, Ukuku, Juneja, & Fan, 2014; Pinheiro, Alegria, Abreu, Gonçalves, & Silva, 2016). However, it is important to mention that these positive effects are seen with specific soft treatments, while high UV doses may have negative effects on texture, color and nutritional characteristics.

Similarly, these higher doses in some cases may affect somehow the texture of the products, since the treatment affects the food structure and cell integrity. In grapes, the presence of nano fractures in skin cells and alteration of cellulose aggregates pattern by UV-C irradiation have been proven (Fava et al., 2011). Also, cell and tissue damage in pods of green bean have been seen (Kasım & Kasım, 2008). In fresh-cut apples, the surface exposed to UV-light showed lost cell integrity (e.g., rupture of the membranes, a decrease in intracellular volume, and loss of turgidity). However, these structural changes have been assessed mainly by optical microscopy observations, and when assessed by instrumental measurements, mechanical property changes were not significant (Gómez, Alzamora, Castro, & Salvatori, 2010; Manzocco et al., 2011) and consumers did not perceive potential changes in texture. Contrarily, in pears, the evidenced changes in structure were detected by consumers and described as a significant loss of hardness and fracturability (Garcia Loredo, Guerrero, & Alzamora, 2013; Schenk, Loredo, Raffellini, Alzamora, & Guerrero, 2012). Surprisingly, these changes did not affect juiciness perception and consumers found treated fruits agreeable.

Lastly, only in scarce studies, the UV-light has been revealed to negatively impact the texture of treated vegetables. For instance, it has been recently assessed that ultraviolet light influences the sensory quality of ginseng roots, resulting in an undesirable softer texture when compared with other preservation treatments (Jin, Huang, Niemira, & Cheng, 2017), although the mechanism of this loss of consistency has not been studied yet. As reported above, UV-light treatments generally help to prolong the fresh firmness of fruits and vegetables. However, observed effects are clearly dose- and product-dependent, confirming that UV treatment must be optimized for each considered food application.

8.2.2.2 Meat Products

The treatment of meat, poultry and fish products by UV light has been mainly studied for surface microbial inactivation and prolonged shelf life of the foods. However, measurements of UV-C effects in texture are less frequent, although several suitable examples can be found in the literature. In this line, also very variable results are reported, depending on the evaluated product.

According to the results published by Oh and coworkers, the combined treatment with UV and chlorine is able to effectively inhibit *Listeria monocytogenes* growth but do not change the texture of chicken breast after one week of storage at 4 °C (Oh, Kang, Oh, & Ha, 2014). Similarly, the texture presented no significant variations between treated and untreated samples also in the case of fresh salmon fillets (Mikš-Krajnik, James Feng, Bang, & Yuk, 2017), processed and cooked meat products (Ha & Kang, 2015; Sommers, Geveke, Pulsfus, & Lemmenes, 2009; Sommers, Scullen, & Sites, 2010), and dried filefish fillets (Park et al., 2014).

However, as it has been mentioned in the previous sections for other products, the optimization of the treatment for each application is crucial here, since very intense UV-light treatments may indirectly alter the texture parameters. These texture damages could be caused by uncontrolled temperature increases in food surface that contribute to moisture loss (Heinrich et al., 2016; Manzocco & Nicoli, 2015).

Contrarily, with proper treatment, the effects of UV-C light in product proteins can be targeted to obtain improvements in texture. Within this context, it has been observed that UV may protect myofibrillar proteins from proteolysis and slow down water holding capacity decrease along the shelf life of sea bass fillets, resulting in a higher hardness and improvement of other textural parameters (Molina, Sáez, Martínez, Guil-Guerrero, & Suárez, 2014), although parallel negative effects were also observed (e.g., decreased collagen content and increased lipid oxidation in fillets).

Also, as previously stated, it is well known that UV irradiation causes the denaturation and cross-linking of certain proteins. This circumstance can be accordingly used for the improvement of strength and elasticity of both meat and fish muscle gels (Ishizaki, Hamada, Iso, & Taguchi, 1993), which present a denser and less porous structure than the untreated samples. Also in fish-gelatin gels, it has been demonstrated by many studies that irradiated samples exhibit significant improvement in the gel texture parameters (Bhat & Karim, 2009; Otoni et al., 2012; Wu, Tsai, & Sung, 2015). Besides, although not directly related to product texture, the functional properties of egg proteins may be also positively impacted by UV light. In this line, foamability and foam stability are improved, and it has been observed an increase on the emulsifying activity in ultraviolet-treated proteins compared to thermal pasteurized samples (de Souza et al., 2015; Kuan, Bhat, & Karim, 2011). In conclusion, taken all these results together, it can be confirmed the potential of employing UV radiation as an alternative method over conventional alternatives for maintaining or improving some of the quality attributes of food products.

8.2.2.3 Dairy Products

As occurs with previously evaluated food products, UV treatment may affect dairy products in such different ways depending on the product and intensity of the treatment.

In milk, UV may cause color and texture changes and other sensorial defects such as off-odors, due to the oxidation of proteins and unsaturated lipids (Orlowska et al., 2012; Rossitto et al., 2012). However, this modification of milk proteins has been demonstrated not to have a negative impact in their coagulation properties and rennet formation (Scheidegger, Pecora, Radici, & Kivatinitz, 2010), and consequently, the resulting cheese made with UV-pasteurized milk should not be affected in terms of texture.

In a positive sense, regarding other properties of milk proteins, UV radiation may lead to enhanced emulsifying and foaming properties of high valued milk-derived ingredients, such as caseins (Kuan et al., 2011, 2013), as also occurred with other food proteins. In solid dairy products, UV light has been studied for surface microbial decontamination and to prolong cheese shelf life. In this case, due to the low transmittance of UV-C light inside the product, this treatment does not induce changes in texture and surface appearance (Lacivita et al., 2016).

8.2.3 Sensorial Properties

Evidence in the compiled scientific literature has perfectly demonstrated that UV light is a promising and viable alternative to thermal pasteurization from a microbiological point of view (Koutchma, 2009). Apart for consideration of UV impact in food safety and nutritional quality, the sensory aspects of treated products must be also evaluated, since consumers' subjective perception of freshness and quality is a key aspect for the purchase decision, and consequently, for the industrial acceptance and implementation of a novel technology.

8.2.3.1 Fruits and Vegetables

In general, UV light preserves the original color and natural properties of a wide variety of fruit juices (Gautam et al., 2017; Taze et al., 2015) and produce minimal or no changes in flavor, essential nutrients and vitamins (Aguilar et al., 2016; Gómez-López et al., 2012). However, as happen with other quality aspects already revisited, high UV doses required in translucent or opaque products to reach an acceptable microbial reduction can adversely change the product's attributes, especially regarding color and the appearance of oxidized flavor, resulting in significant quality changes and reduced consumer's acceptability (Bermúdez-Aguirre & Barbosa-Cánovas, 2013; Caminiti et al., 2012; Santhirasegaram et al., 2015). These changes in color are sometimes useful by degrading undesirable brown colors formed in the fruit derivatives (Ibarz, Pagán, Panadés, & Garza, 2005). Also, in a positive way, the treatment by UV light can be beneficial, regarding the inhibition of enzymatic activity whose reactions can lead to quality losses during storage: UV light has been proved to inactivate oxidases and pectin methylesterase responsible for browning and clarification of fruit juices and other fruit derivatives (Aguilar et al., 2016; Falguera, Pagán, & Ibarz, 2011; Manzocco, Dri, & Quarta, 2009; Manzocco, Quarta, & Dri, 2009; Müller, Noack, Greiner, Stahl, & Posten, 2014; Tran & Farid, 2004).

In solid fruits, some differences are noted between whole or cut fruits. In this line, in whole fruits, no positive or negative influence of UV-C treatment quality can be generally found (Marquenie, 2002; Marquenie et al., 2002), except for slight changes in color (Birmpa et al., 2013) or even a positive influence by delaying ripening (Pinheiro et al., 2016).

Regarding fresh-cut fruits, UV light may produce slight direct degradation of fruit color (Romero, Colivet, Aron, & Ramosvillarroel, 2017). These color changes are observed also in other vegetal products such as a significant increase in the lightness of peanut oil (Shen et al., 2014). However, this treatment can be also used for controlling enzymatic browning (Lante, Tinello, & Nicoletto, 2016; Moreno et al., 2017; Şakiroğlu, Birdal, Başlar, & Öztürk, 2016). Apart from enzymatic inactivation, it has been evidenced the formation of a thin, dried film on the surface of the products due to the UV treatment, that seem to protect cut fruit from oxidation, allowing the retention of fresh-like appearance and flavor during storage (Manzocco et al., 2011). However, as cited in previous sections, this damage in the structure of cells due to intensive ultraviolet exposure can also produce the contrary effects to those positive ones previously mentioned (Koutchma & Orlowska, 2012). Thus, although these changes were not significant in global texture, surface and color changes may occur, such as accelerated peel and surface browning in citrus and pineapples (Ben-Yehoshua, Rodov, Kim, & Carmeli, 1992; Pan & Zu, 2012) and moisture loss (Manzocco & Nicoli, 2015). Nevertheless, the potential appearance of negative effects at high doses opens to the possibility of decreasing the treatment intensity by combining two or more treatments in order to preserve the fruit and vegetable quality without decreasing the microbial and enzymatic inactivation properties (Marquenie, 2002; Şakiroğlu et al., 2016; Sampedro & Fan, 2014).

8.2.3.2 Meat Products

Meat color is considered one of the most influential factors in consumer purchasing decisions. Although references in the literature are less numerous than in fruit products, the use of UV light to reduce surface microbial contamination has been normally reported not to have significant detrimental effect in the color of meat and meat products, mainly at moderate dosage or in combined treatments (Ha & Kang, 2015; Koutchma et al., 2009; Lázaro et al., 2014; Yang et al., 2017).

However, concerning products containing sensitive pigments such as salmon fillets, sensory results showed that the treatments involving UV irradiation gave significantly lower color scores (Mikš-Krajnik et al., 2017). Once again, higher doses of UV light may jeopardize sensory characteristics of treated products: color alterations and fat oxidation are the most remarkable changes caused by intense treatments, decreasing overall acceptability (Cooper et al., 2016; Park et al., 2014; Park & Ha, 2015). Indirectly, the potential excessive temperature increase in food surface can also contribute to quality changes in the surface of meat and fish (Heinrich et al., 2016; Ozer & Demirci, 2006).

In egg products, UV-C treatment has been shown to be effective in reducing microbial load in white egg, also improving its properties. Contrarily, in egg yolk and whole egg, it may cause browning and lipid oxidation, reducing their sensory quality (de Souza & Fernández, 2011; Manzocco, Panozzo, & Nicoli, 2012).

8.2.3.3 Dairy Products

The dairy industry is the fastest growing agricultural sector in the world, so there is a need to develop novel processing techniques, an alternative to thermal processes to meet this global demand, and offer consumers high-quality milk and dairy products (Datta et al., 2015).

As said before, UV treatment causes sensory defects such as color changes and presence of off-odors in milk that are caused by oxidation of proteins and lipids (Rossitto et al., 2012; Scheidegger et al., 2010). In addition, although an increase in vitamin D can be obtained at short UV wavelength, the reduction in the content of other important vitamins can lead to an overall decrease of milk quality (Guneser & Karagul Yuceer, 2012). These sensory changes have also been observed in goat milk (Matak et al., 2007), deducing that milk components are extremely light-sensitive products (Koutchma, 2009). However, these organoleptic changes have not been considered as a safety concern by European administration regarding the approval of UV-light for milk treatment (EFSA, 2016). Anyway, trying to minimize the appearance of sensory defects in UV-light treated milk products means a major challenge for future research and UV application in the dairy industry. In solid dairy products, scarce studies have been published, but in general, the application of UV light in cheese do not promote changes in terms of color, texture and surface appearance (Lacivita et al., 2016).

8.3 Conclusions

As the global food supply moves toward fresh and minimally processed products, the industry continues to seek new technologies to control possible microbiological hazards while maintaining food quality characteristics. In this context, UV-light has been stipulated as a promising technology to provide the quality attributes demanded by consumers and to ensure the expected food safety.

The major advantages of UV are mainly summarized as the extensive applicability to most types of microorganisms, the lower cost, and the convenient manipulation. Also, the exposure to UV-light could be an advantage regarding food quality in most cases. However, the most significant and critical challenges to the commercial application of UV-C technology are its limited penetration power and the need for achieving dose uniformity in food products. Also, more intense studies of the effect of UV radiation on the structure and functional properties of light-sensitive food components, as well as protecting certain nutrients from photodegradation is still a major challenge to take over by food scientist in order to impulse its application in food industry. The development of combined processes with conventional and other emerging preservation methods at a moderate intensity that potentate UV lethal effect, seem to be an interesting solution to this threat. In any case, the loss of nutrients by UV light seems to be lower than with thermal treatment, so regarding other parameters (e.g., safety issues, food characteristics, economic issues, etc.) and the continuous development of improved UV sources and processing systems for specific foods, it is clear that UV- radiation still shows a great potential: As seen in this chapter, fruit and vegetable applications seem to be still a majority, but novel applications in opaque and solid foods are also being studied and are on the rise, and will contribute to a more extended commercial use of UV-based treatment for pasteurization foods.

References

Aguilar, K., Ibarz, R., Garvín, A., & Ibarz, A. (2016). Effect of UV–Vis irradiation on enzymatic activities and the physicochemical properties of nectarine juices from different varieties. *LWT - Food Science and Technology, 65*, 969–977.

Anugu, A. K., Yang, W., Shriver, S., Chung, S. Y., & Percival, S. S. (2010). Efficacy of pulsed ultraviolet light in reducing the allergenicity of isolated egg proteins. In *Institute of Food Technologists*. Chicago, IL.

Arroyo, C., Dorozko, A., Gaston, E., O'Sullivan, M., Whyte, P., & Lyng, J. G. (2017). Light based technologies for microbial inactivation of liquids, bead surfaces and powdered infant formula. *Food Microbiology, 67*, 49–57.

Augusto, P. E. D., Ibarz, R., Garvín, A., & Ibarz, A. (2015). Peroxidase (POD) and polyphenol oxidase (PPO) photo-inactivation in a coconut water model solution using ultraviolet (UV). *Food Research International (Ottawa, Ont.), 74*, 151–159.

Baron, C. P., Børresen, T., & Jacobsen, C. (2005). UV treatment of fishmeal: A method to remove dioxins. *Journal of Agricultural and Food Chemistry, 53*(18), 7091–7097.

Basaran, N., Quintero-Ramos, A., Moake, M. M., Churey, J. J., & Worobo, R. W. (2004). Influence of apple cultivars on inactivation of different strains of *Escherichia coli* O157:H7 in apple cider by UV irradiation. *Applied and Environmental Microbiology, 70*(10), 6061–6065.

Ben-Yehoshua, S., Rodov, V., Kim, J. J., & Carmeli, S. (1992). Preformed and induced antifungal materials of citrus fruits in relation to the enhancement of decay resistance by heat and ultraviolet treatments. *Journal of Agricultural and Food Chemistry, 40*(7), 1217–1221.

Bermúdez-Aguirre, D., & Barbosa-Cánovas, G. V. (2013). Disinfection of selected vegetables under nonthermal treatments: Chlorine, acid citric, ultraviolet light and ozone. *Food Control, 29*(1), 82–90.

Bhat, R., & Karim, A. A. (2009). Ultraviolet irradiation improves gel strength of fish gelatin. *Food Chemistry, 113*(4), 1160–1164.

Bintsis, T., Litopoulou-Tzanetaki, E., & Robinson, R. K. (2000). Existing and potential applications of ultraviolet light in the food industry – a critical review. *Journal of the Science of Food and Agriculture, 80*(6), 637–645.

Birmpa, A., Sfika, V., & Vantarakis, A. (2013). Ultraviolet light and Ultrasound as non-thermal treatments for the inactivation of microorganisms in fresh ready-to-eat foods. *International Journal of Food Microbiology, 167*(1), 96–102.

Bravo, S., García-Alonso, J., Martín-Pozuelo, G., Gómez, V., García-Valverde, V., Navarro-González, I., & Periago, M. J. (2013). Effects of postharvest UV-C treatment on carotenoids and phenolic compounds of vine-ripe tomatoes. *International Journal of Food Science & Technology, 48*(8), 1744–1749.

Caminiti, I. M., Palgan, I., Muñoz, A., Noci, F., Whyte, P., Morgan, D. J., … Lyng, J. G. (2012). The effect of ultraviolet light on microbial inactivation and quality attributes of apple juice. *Food and Bioprocess Technology, 5*(2), 680–686.

Cantos, E., Espín, J. C., & Tomás-Barberán, F. A. (2001). Postharvest induction modeling method using UV irradiation pulses for obtaining resveratrol-enriched table grapes: A new "functional" fruit? *Journal of Agricultural and Food Chemistry, 49*(10), 5052–5058.

Cappozzo, J. C., Koutchma, T., & Barnes, G. (2015). Chemical characterization of milk after treatment with thermal (HTST and UHT) and nonthermal (turbulent flow ultraviolet) processing technologies. *Journal of Dairy Science, 98*(8), 5068–5079.

Chen, C., Hu, W., He, Y., Jiang, A., & Zhang, R. (2016). Effect of citric acid combined with UV-C on the quality of fresh-cut apples. *Postharvest Biology and Technology, 111*, 126–131.

Choi, I.-L., Yoo, T. J., & Kang, H.-M. (2015). UV-C treatments enhance antioxidant activity, retain quality and microbial safety of fresh-cut paprika in MA storage. *Horticulture, Environment, and Biotechnology, 56*(3), 324–329.

Chung, S.-Y., Yang, W., & Krishnamurthy, K. (2008). Effects of pulsed UV-light on peanut allergens in extracts and liquid peanut butter. *Journal of Food Science, 73*(5), C400–C404.

Cooper, J. V., Wiegand, B. R., Koc, A. B., Schumacher, L., Grün, I., & Lorenzen, C. L. (2016). RAPID COMMUNICATION: Impact of contemporary light sources on oxidation of fresh ground beef. *Journal of Animal Science, 94*(10), 4457–4462.

Datta, N., Harimurugan, P., & Palombo, E. A. (2015). Ultraviolet and pulsed light technologies in dairy processing. In N. Datta & P. M. Tomasula (Eds.), *Emerging dairy processing technologies* (pp. 181–204). Wiley-Blackwell. Chichester, West Sussex, UK.

de Oliveira Bottino, F., Rodrigues, B. L., de Nunes Ribeiro, J. D., Lázaro, C. A., & Conte-Junior, C. A. (2016). Effect of UV-C radiation on shelf life of vacuum package *Colossoma macropomum* x *Piaractus mesopotamicus* fillets. *Procedia Food Science, 7*, 13–16.

de Souza, P. M., & Fernández, A. (2011). Effects of UV-C on physicochemical quality attributes and *Salmonella enteritidis* inactivation in liquid egg products. *Food Control, 22*(8), 1385–1392.

de Souza, P. M., Müller, A., Beniaich, A., Mayer-Miebach, E., Oehlke, K., Stahl, M., … Fernández, A. (2015). Functional properties and nutritional composition of liquid egg products treated in a coiled tube UV-C reactor. *Innovative Food Science & Emerging Technologies, 32*, 156–164.

Diao, E., Li, X., Zhang, Z., Ma, W., Ji, N., & Dong, H. (2015). Ultraviolet irradiation detoxification of aflatoxins. *Trends in Food Science & Technology, 42*(1), 64–69.

Díaz, O., Candia, D., & Cobos, Á. (2017). Whey protein film properties as affected by ultraviolet treatment under alkaline conditions. *International Dairy Journal, 73*, 84–91.

Du, W.-X., Avena-Bustillos, R. J., Breksa, A. P., & McHugh, T. H. (2012). Effect of UV-B light and different cutting styles on antioxidant enhancement of commercial fresh-cut carrot products. *Food Chemistry, 134*(4), 1862–1869.

Du, W.-X., Avena-Bustillos, R. J., Breksa, A. P., & McHugh, T. H. (2014). UV-B light as a factor affecting total soluble phenolic contents of various whole and fresh-cut specialty crops. *Postharvest Biology and Technology, 93*, 72–82.

EFSA. (2005). Opinion of the scientific panel on biological hazards on the request from the commission related to the microbiological risks on washing of table eggs. *The EFSA Journal, 269*, 1–39. https://doi.org/10.2903/j.efsa.2005.269

EFSA. (2015). Scientific opinion on the safety of UV-treated bread as a novel food. *The EFSA Journal, 13*(7), 4148–4164. https://doi.org/10.2903/j.efsa.2015.4148

EFSA. (2016). Safety of UV-treated milk as a novel food pursuant to Regulation (EC) No 258/97. *The EFSA Journal, 14*(1), 4370–4384. https://doi.org/10.2903/j.efsa.2016.4370

European Commission. (1997). *Regulation EC No. 258/97 of the European Parliament and of the Council concerning novel foods and novel food ingredients.* (No. Official Journal L 043) (pp. 1–9). European Commission. Retrieved from https://www.ecolex.org/details/legislation/regulation-ec-no-25897-of-the-european-parliament-and-of-the-council-concerning-novel-foods-and-novel-food-ingredients-lex-faoc018638/

Falguera, V., Pagán, J., & Ibarz, A. (2011). Effect of UV irradiation on enzymatic activities and physicochemical properties of apple juices from different varieties. *LWT - Food Science and Technology, 44*(1), 115–119.

Fan, X., Huang, R., & Chen, H. (2017). Application of ultraviolet C technology for surface decontamination of fresh produce. *Trends in Food Science & Technology, 70*, 9–19.

Fava, J., Hodara, K., Nieto, A., Guerrero, S., Alzamora, S. M., & Castro, M. A. (2011). Structure (micro, ultra, nano), color and mechanical properties of *Vitis labrusca* L. (grape berry) fruits treated by hydrogen peroxide, UV–C irradiation and ultrasound. *Food Research International, 44*(9), 2938–2948.

Fiocchi, A., Schünemann, H. J., Brozek, J., Restani, P., Beyer, K., Troncone, R., … Lockey, R. F. (2010). Diagnosis and rationale for action against cow's milk allergy (DRACMA): A summary report. *The Journal of Allergy and Clinical Immunology, 126*(6), 1119–1128.e12.

Fonseca, J. M., & Rushing, J. W. (2006). Effect of ultraviolet-C light on quality and microbial population of fresh-cut watermelon. *Postharvest Biology and Technology, 40*(3), 256–261.

Formica-Oliveira, A. C., Martínez-Hernández, G. B., Díaz-López, V., Artés, F., & Artés-Hernández, F. (2017). Effects of UV-B and UV-C combination on phenolic compounds biosynthesis in fresh-cut carrots. *Postharvest Biology and Technology, 127*, 99–104.

García Carrillo, M., Ferrario, M., & Guerrero, S. (2017). Study of the inactivation of some microorganisms in turbid carrot-orange juice blend processed by ultraviolet light assisted by mild heat treatment. *Journal of Food Engineering, 212*, 213–225.

Garcia Loredo, A. B., Guerrero, S. N., & Alzamora, S. M. (2013). Impact of combined ascorbic acid/$CaCl_2$, hydrogen peroxide and ultraviolet light treatments on structure, rheological properties and texture of fresh-cut pear (William var.). *Journal of Food Engineering, 114*(2), 164–173.

Gautam, D., Umagiliyage, A. L., Dhital, R., Joshi, P., Watson, D. G., Fisher, D. J., & Choudhary, R. (2017). Nonthermal pasteurization of tender coconut water using a continuous flow coiled UV reactor. *LWT - Food Science and Technology, 83*, 127–131.

Gayán, E., Condón, S., & Álvarez, I. (2014). Biological aspects in food preservation by ultraviolet light: A review. *Food and Bioprocess Technology, 7*(1), 1–20.

George, D. S., Razali, Z., Santhirasegaram, V., & Somasundram, C. (2015). Effects of ultraviolet light (UV-C) and heat treatment on the quality of fresh-cut Chokanan mango and Josephine pineapple. *Journal of Food Science, 80*(2), S426–S434.

George, D. S., Razali, Z., Santhirasegaram, V., & Somasundram, C. (2016). Effect of postharvest ultraviolet-C treatment on the proteome changes in fresh cut mango (*Mangifera indica* L. cv. Chokanan). *Journal of the Science of Food and Agriculture, 96*(8), 2851–2860.

Gómez, P. L., Alzamora, S. M., Castro, M. A., & Salvatori, D. M. (2010). Effect of ultraviolet-C light dose on quality of cut-apple: Microorganism, color and compression behavior. *Journal of Food Engineering, 98*(1), 60–70.

Gómez-López, V. M., Koutchma, T., & Linden, K. (2012). Ultraviolet and pulsed light processing of fluid foods. In P. J. Cullen, B. K. Tiwari, & V. P. Valdramidis (Eds.), *Novel thermal and non-thermal technologies for fluid foods* (pp. 185–223). San Diego, CA: Academic.

Gouma, M., Álvarez, I., Condón, S., & Gayán, E. (2015). Modelling microbial inactivation kinetics of combined UV-H treatments in apple juice. *Innovative Food Science & Emerging Technologies, 27*, 111–120.

Guerrero-Beltrán, J. A., & Barbosa-Cánovas, G. V. (2004). Review: Advantages and limitations on processing foods by UV light. *Food Science and Technology International, 10*(3), 137–147.

Guneser, O., & Karagul Yuceer, Y. (2012). Effect of ultraviolet light on water- and fat-soluble vitamins in cow and goat milk. *Journal of Dairy Science, 95*(11), 6230–6241.

Gutiérrez, D. R., Chaves, A. R., & Rodríguez, S. D. C. (2017). Use of UV-C and gaseous ozone as sanitizing agents for keeping the quality of fresh-cut rocket *(Eruca sativa* mill). *Journal of Food Processing and Preservation, 41*(3), e12968. https://doi.org/10.1111/jfpp.12968

Ha, J.-W., & Kang, D.-H. (2015). Enhanced inactivation of food-borne pathogens in ready-to-eat sliced ham by near-infrared heating combined with UV-C irradiation and mechanism of the synergistic bactericidal action. *Applied and Environmental Microbiology, 81*(1), 2–8.

Heinrich, V., Zunabovic, M., Varzakas, T., Bergmair, J., & Kneifel, W. (2016). Pulsed light treatment of different food types with a special focus on meat: A critical review. *Critical Reviews in Food Science and Nutrition, 56*(4), 591–613.

Hu, G., Zheng, Y., Liu, Z., Xiao, Y., Deng, Y., & Zhao, Y. (2017). Effects of high hydrostatic pressure, ultraviolet light-C, and far-infrared treatments on the digestibility, antioxidant and antihypertensive activity of α-casein. *Food Chemistry, 221*, 1860–1866.

Ibarz, A., Pagán, J., Panadés, R., & Garza, S. (2005). Photochemical destruction of color compounds in fruit juices. *Journal of Food Engineering, 69*(2), 155–160.

Ishizaki, S., Hamada, M., Iso, N., & Taguchi, T. (1993). Effect of ultraviolet irradiation on rheological properties of thermal gels from sardine and pork meat pastes. *Nippon Suisan Gakkaishi, 59*, 1219–1224.

Islam, M. S., Patras, A., Pokharel, B., Vergne, M. J., Sasges, M., Begum, A., … Xiao, H. (2016). Effect of UV irradiation on the nutritional quality and cytotoxicity of apple juice. *Journal of Agricultural and Food Chemistry, 64*(41), 7812–7822.

Jermann, C., Koutchma, T., Margas, E., Leadley, C., & Ros-Polski, V. (2015). Mapping trends in novel and emerging food processing technologies around the world. *Innovative Food Science & Emerging Technologies, 31*, 14–27.

Jiang, T., Jahangir, M. M., Jiang, Z., Lu, X., & Ying, T. (2010). Influence of UV-C treatment on antioxidant capacity, antioxidant enzyme activity and texture of postharvest shiitake (*Lentinus edodes*) mushrooms during storage. *Postharvest Biology and Technology, 56*(3), 209–215.

Jin, T. Z., Huang, M., Niemira, B. A., & Cheng, L. (2017). Microbial reduction and sensory quality preservation of fresh ginseng roots using nonthermal processing and antimicrobial packaging. *Journal of Food Processing and Preservation, 41*(1), e12871. https://doi.org/10.1111/jfpp.12871

Kanazawa, K., Hashimoto, T., Yoshida, S., Sungwon, P., & Fukuda, S. (2012). Short photoirradiation induces flavonoid synthesis and increases its production in postharvest vegetables. *Journal of Agricultural and Food Chemistry, 60*(17), 4359–4368.

Kasım, M. U., & Kasım, R. (2008). Double-sided UV-C treatments delayed chlorophyll degradation but increased chilling injury of green bean (*Phaseolus vulgaris* L. cv. Helda) during storage. *Journal of Food, Agriculture & Environment, 6*, 176–180.

Kershner, K. (2015). The role of UV in industry. *Water and Wastes Digest, 20*.

Kim, S. K., Bae, R. N., & Chun, C. (2011). Changes in bioactive compounds contents of "Maehyang" and "Seolhyang" strawberry fruits by UV light illumination. *Korean Journal of Horticultural Science & Technology, 29*(3), 172–180.

Kim, Y.-H., Jeong, S.-G., Back, K.-H., Park, K.-H., Chung, M.-S., & Kang, D.-H. (2013). Effect of various conditions on inactivation of *Escherichia coli* O157:H7, *Salmonella* Typhimurium, and *Listeria monocytogenes* in fresh-cut lettuce using ultraviolet radiation. *International Journal of Food Microbiology, 166*(3), 349–355.

Koutchma, T. (2008). UV light for processing foods. *Ozone: Science & Engineering, 30*(1), 93–98.

Koutchma, T. (2009). Advances in ultraviolet light technology for non-thermal processing of liquid foods. *Food and Bioprocess Technology, 2*(2), 138–155.

Koutchma, T., Forney, L. J., & Moraru, C. I. (2009). *Ultraviolet light in food technology: Principles and applications*. Boca Raton, FL: CRC Press.

Koutchma, T., Keller, S., Chirtel, S., & Parisi, B. (2004). Ultraviolet disinfection of juice products in laminar and turbulent flow reactors. *Innovative Food Science & Emerging Technologies, 5*(2), 179–189.

Koutchma, T., & Orlowska, M. (2012). Ultraviolet light for processing fruits and fruit products. In S. Fernandes & F. A. N. Fernandes (Eds.), *Advances in fruit processing technologies*. Boca Raton, FL: CRC Press/Taylor and Francis.

Koutchma, T., & Parisi, B. P. (2004). Biodosimetry inactivation in of *Escherichia coli* UV model juices with regard to dose distribution in annular UV reactors. *Journal of Food Science, 69*(1), FEP14–FEP22.

Koutchma, T., Popović, V., Ros-Polski, V., & Popielarz, A. (2016). Effects of ultraviolet light and high-pressure processing on quality and health-related constituents of fresh juice products. *Comprehensive Reviews in Food Science and Food Safety, 15*(5), 844–867.

Kuan, Y.-H., Bhat, R., & Karim, A. A. (2011). Emulsifying and foaming properties of ultraviolet-irradiated egg white protein and sodium caseinate. *Journal of Agricultural and Food Chemistry, 59*(8), 4111–4118.

Kuan, Y.-H., Bhat, R., Patras, A., & Karim, A. A. (2013). Radiation processing of food proteins – A review on the recent developments. *Trends in Food Science & Technology, 30*(2), 105–120.

Lacivita, V., Conte, A., Manzocco, L., Plazzotta, S., Zambrini, V. A., Del Nobile, M. A., & Nicoli, M. C. (2016). Surface UV-C light treatments to prolong the shelf-life of Fiordilatte cheese. *Innovative Food Science & Emerging Technologies, 36*, 150–155.

Lamikanra, O., Kueneman, D., Ukuku, D., & Bett-Garber, K. L. (2005). Effect of processing under ultraviolet light on the shelf life of fresh-cut cantaloupe melon. *Journal of Food Science, 70*(9), C534–C539.

Lante, A., Tinello, F., & Nicoletto, M. (2016). UV-A light treatment for controlling enzymatic browning of fresh-cut fruits. *Innovative Food Science & Emerging Technologies, 34*, 141–147.

Lavilla, M., & Gayán, E. (2018). Consumer acceptance and marketing of foods processed through emerging technologies. In F. J. Barba, A. S. Sant'Ana, V. Orlien, & M. Koubaa (Eds.), *Innovative technologies for food preservation: Inactivation of spoilage and pathogenic microorganisms* (pp. 233–253). Academis Press/Elsevier. London, UK.

Lázaro, C. A., Conte-Júnior, C. A., Monteiro, M. L. G., Canto, A. C. V. S., Costa-Lima, B. R. C., Mano, S. B., & Franco, R. M. (2014). Effects of ultraviolet light on biogenic amines and other quality indicators of chicken meat during refrigerated storage. *Poultry Science, 93*(9), 2304–2313.

Liu, J., Stevens, C., Khan, V. A., Lu, J. Y., Wilson, C. L., Adeyeye, O., … Droby, S. (1993). Application of ultraviolet-C light on storage rots and ripening of tomatoes. *Journal of Food Protection, 56*(10), 868–873.

Longo, E., Morozova, K., & Scampicchio, M. (2017). Effect of light irradiation on the antioxidant stability of oleuropein. *Food Chemistry, 237*, 91–97.

Ma, L., Zhang, M., Bhandari, B., & Gao, Z. (2017). Recent developments in novel shelf life extension technologies of fresh-cut fruits and vegetables. *Trends in Food Science & Technology, 64*, 23–38.

Manzocco, L., Da Pieve, S., Bertolini, A., Bartolomeoli, I., Maifreni, M., Vianello, A., & Nicoli, M. C. (2011). Surface decontamination of fresh-cut apple by UV-C light exposure: Effects on structure, colour and sensory properties. *Postharvest Biology and Technology, 61*(2), 165–171.

Manzocco, L., Dri, A., & Quarta, B. (2009). Inactivation of pectic lyases by light exposure in model systems and fresh-cut apple. *Innovative Food Science & Emerging Technologies, 10*(4), 500–505.

Manzocco, L., & Nicoli, M. C. (2015). Surface processing: Existing and potential applications of ultraviolet light. *Critical Reviews in Food Science and Nutrition, 55*(4), 469–484.

Manzocco, L., Panozzo, A., & Nicoli, M. C. (2012). Effect of ultraviolet processing on selected properties of egg white. *Food Chemistry, 135*(2), 522–527.

Manzocco, L., Quarta, B., & Dri, A. (2009). Polyphenoloxidase inactivation by light exposure in model systems and apple derivatives. *Innovative Food Science & Emerging Technologies, 10*(4), 506–511.

Marquenie, D. (2002). *Evaluation of physical techniques for surface disinfection of strawberry and sweet cherry* (Thesis). Katholieke Universiteit Leuven, Leuven, Belgium.

Marquenie, D., Michiels, C. W., Geeraerd, A. H., Schenk, A., Soontjens, C., Van Impe, J. F., & Nicolaï, B. M. (2002). Using survival analysis to investigate the effect of UV-C and heat treatment on storage rot of strawberry and sweet cherry. *International Journal of Food Microbiology, 73*(2), 187–196.

Marquenie, D., Michiels, C. W., Van Impe, J. F., Schrevens, E., & Nicolaï, B. N. (2003). Pulsed white light in combination with UV-C and heat to reduce storage rot of strawberry. *Postharvest Biology and Technology, 28*(3), 455–461.

Martínez-Hernández, G. B., Navarro-Rico, J., Gómez, P. A., Otón, M., Artés, F., & Artés-Hernández, F. (2015). Combined sustainable sanitising treatments to reduce *Escherichia coli* and *Salmonella Enteritidis* growth on fresh-cut kailan-hybrid broccoli. *Food Control*, (47), 312–317.

Matak, K. E., Sumner, S. S., Duncan, S. E., Hovingh, E., Worobo, R. W., Hackney, C. R., & Pierson, M. D. (2007). Effects of ultraviolet irradiation on chemical and sensory properties of goat milk. *Journal of Dairy Science, 90*(7), 3178–3186.

Meng, X., & Chan, W. (2017). Determination of 2-alkylcyclobutanones in ultraviolet light-irradiated fatty acids, triglycerides, corn oil, and pork samples: Identifying a new source of 2-alkylcyclobutanones. *Food Chemistry, 217*, 352–359.

Mikš-Krajnik, M., James Feng, L. X., Bang, W. S., & Yuk, H.-G. (2017). Inactivation of *Listeria monocytogenes* and natural microbiota on raw salmon fillets using acidic electrolyzed water, ultraviolet light or/and ultrasounds. *Food Control, 74*, 54–60.

Misra, N. N. (2015). The contribution of non-thermal and advanced oxidation technologies towards dissipation of pesticide residues. *Trends in Food Science & Technology, 45*(2), 229–244.

Molina, B., Sáez, M. I., Martínez, T. F., Guil-Guerrero, J. L., & Suárez, M. D. (2014). Effect of ultraviolet light treatment on microbial contamination, some textural and organoleptic parameters of cultured sea bass fillets (*Dicentrarchus labrax*). *Innovative Food Science & Emerging Technologies, 26*, 205–213.

Monteiro, M. L. G., Mársico, E. T., da Canto, A. C. V. C. S., da Costa-Lima, B. R. C., da Costa, M. P., Viana, F. M., … Conte-Junior, C. A. (2017). Impact of UV-C light on the fatty acid profile and oxidative stability of Nile Tilapia (*Oreochromis niloticus*) fillets. *Journal of Food Science, 82*(4), 1028–1036.

Moreno, C., Andrade-Cuvi, M. J., Zaro, M. J., Darre, M., Vicente, A. R., & Concellón, A. (2017). Short UV-C treatment prevents browning and extends the shelf-life of fresh-cut carambola. *Journal of Food Quality*. https://doi.org/10.1155/2017/2548791

Mori, M., Hamamoto, A., Takahashi, A., Nakano, M., Wakikawa, N., Tachibana, S., … Kinouchi, Y. (2007). Development of a new water sterilization device with a 365 nm UV-LED. *Medical & Biological Engineering & Computing, 45*(12), 1237–1241.

Mukhopadhyay, S., Ukuku, D. O., Juneja, V., & Fan, X. (2014). Effects of UV-C treatment on inactivation of *Salmonella enterica* and *Escherichia coli* O157:H7 on grape tomato surface and stem scars, microbial loads, and quality. *Food Control, 44*, 110–117.

Müller, A., Noack, L., Greiner, R., Stahl, M. R., & Posten, C. (2014). Effect of UV-C and UV-B treatment on polyphenol oxidase activity and shelf life of apple and grape juices. *Innovative Food Science & Emerging Technologies, 26*, 498–504.

Nogales-Delgado, S., Fernández-León, A. M., Delgado-Adámez, J., Hernández-Méndez, M. T., & Bohoyo-Gil, D. (2014). Lactic acid and ultraviolet-C as sanitizer for preserving quality of minimally processed romaine lettuce. *Journal of Food Processing and Preservation, 38*(3), 774–783.

Oh, S. R., Kang, I., Oh, M. H., & Ha, S. D. (2014). Inhibitory effect of chlorine and ultraviolet radiation on growth of *Listeria monocytogenes* in chicken breast and development of predictive growth models. *Poultry Science, 93*(1), 200–207.
Ohlsson, T. (2002). Minimal processing of food with non-thermal methods. In T. Ohlsson & N. Bengtsson (Eds.), *Minimal processing technologies in the food industry* (pp. 34–60). Boca Raton, FL: CRC Press.
Orlowska, M., Koutchma, T., Grapperhaus, M., Gallagher, J., Schaefer, R., & Defelice, C. (2012). Continuous and pulsed ultraviolet light for nonthermal treatment of liquid foods. Part 1: Effects on quality of fructose solution, apple juice, and milk. *Food and Bioprocess Technology, 6*(6), 1580–1592.
Otoni, C. G., Avena-Bustillos, R. J., Chiou, B.-S., Bilbao-Sainz, C., Bechtel, P. J., & McHugh, T. H. (2012). Ultraviolet-B radiation induced cross-linking improves physical properties of cold- and warm-water fish gelatin gels and films. *Journal of Food Science, 77*(9), E215–E223.
Ozen, B. F., & Floros, J. D. (2001). Effects of emerging food processing techniques on the packaging materials. *Trends in Food Science & Technology, 12*(2), 60–67.
Ozer, N. P., & Demirci, A. (2006). Inactivation of *Escherichia coli* O157:H7 and *Listeria monocytogenes* inoculated on raw salmon fillets by pulsed UV-light treatment. *International Journal of Food Science & Technology, 41*(4), 354–360.
Pala, Ç. U., & Toklucu, A. K. (2013). Effects of UV-C light processing on some quality characteristics of grape juices. *Food and Bioprocess Technology, 6*(3), 719–725.
Pan, Y. G., & Zu, H. (2012). Effect of UV-C radiation on the quality of fresh-cut pineapples. *Procedia Engineering, 37*, 113–119.
Park, S. K., Cho, J. M., & Rhee, C. O. (2003). Effects of ultraviolet radiation on the red colored film of soy protein isolate. *Food Science and Biotechnology, 12*, 385–389.
Park, S. Y., & Ha, S.-D. (2015). Ultraviolet-C radiation on the fresh chicken Breast: Inactivation of major foodborne viruses and changes in physicochemical and sensory qualities of product. *Food and Bioprocess Technology, 8*(4), 895–906.
Park, S. Y., Lee, N.-Y., Kim, S.-H., Cho, J.-I., Lee, H.-J., & Ha, S.-D. (2014). Effect of ultraviolet radiation on the reduction of major food spoilage molds and sensory quality of the surface of dried filefish (*Stephanolepis cirrhifer*) fillets. *Food Research International, 62*, 1108–1112.
Pellicer, J. A., & Gómez-López, V. M. (2017). Pulsed light inactivation of horseradish peroxidase and associated structural changes. *Food Chemistry, 237*, 632–637.
Pereira, R. N., & Vicente, A. A. (2010). Environmental impact of novel thermal and non-thermal technologies in food processing. *Food Research International, 43*(7), 1936–1943.
Pinheiro, J. C., Alegria, C. S. M., Abreu, M. M. M. N., Gonçalves, E. M., & Silva, C. L. M. (2016). Evaluation of alternative preservation treatments (water heat treatment, ultrasounds, thermosonication and UV-C radiation) to improve safety and quality of whole tomato. *Food and Bioprocess Technology, 9*(6), 924–935.
Promyou, S., & Supapvanich, S. (2016). Physicochemical changes in 'kaew kamin' mango fruit illuminated with ultraviolet-C (UV-C) during storage. *Journal of Agricultural Science and Technology, 18*(1), 145–154.
Raybaudi-Massilia, R., Calderón-Gabaldón, M. I., Mosqueda-Melgar, J., & Tapia, M. S. (2013). Inactivation of *Salmonella enterica* ser. Poona and *Listeria monocytogenes* on fresh-cut "Maradol" red papaya (*Carica papaya* L) treated with UV-C light and malic acid. *Journal für Verbraucherschutz und Lebensmittelsicherheit, 8*(1–2), 37–44.
Rhim, J. W., Gennadios, A., Fu, D., Weller, C. L., & Hanna, M. A. (1999). Properties of ultraviolet irradiated protein films. *LWT - Food Science and Technology, 32*(3), 129–133.
Rodoni, L. M., Zaro, M. J., Hasperué, J. H., Concellón, A., & Vicente, A. R. (2015). UV-C treatments extend the shelf life of fresh-cut peppers by delaying pectin solubilization and inducing local accumulation of phenolics. *LWT - Food Science and Technology, 63*(1), 408–414.
Rodriguez-Gonzalez, O., Buckow, R., Koutchma, T., & Balasubramaniam, V. M. (2015). Energy requirements for alternative food processing technologies-principles, assumptions, and evaluation of efficiency. *Comprehensive Reviews in Food Science and Food Safety, 14*(5), 536–554.

Romero, L., Colivet, J., Aron, N. M., & Ramosvillarroel, A. (2017). Impact of ultraviolet light on quality attributes of stored fresh-cut mango. *The Annals of the University Dunarea de Jos of Galati Fascicle VI – Food Technology, 41*, 62–80.

Rossitto, P. V., Cullor, J. S., Crook, J., Parko, J., Sechi, P., & Cenci-Goga, B. T. (2012). Effects of UV irradiation in a continuous turbulent flow UV reactor on microbiological and sensory characteristics of cow's milk. *Journal of Food Protection, 75*(12), 2197–2207.

Şakiroğlu, H., Birdal, C., Başlar, M., & Öztürk, A. E. (2016). Inactivation kinetics of polyphenol oxidase in an aqueous model system under stand-alone and combined ultrasound and ultraviolet treatments. *International Journal of Food Properties, 19*(7), 1535–1543.

Sales, J. M., & Resurreccion, A. V. A. (2010). Phenolic profile, antioxidants, and sensory acceptance of bioactive-enhanced peanuts using ultrasound and UV. *Food Chemistry, 122*(3), 795–803.

Sampedro, F., & Fan, X. (2014). Inactivation kinetics and photoreactivation of vegetable oxidative enzymes after combined UV-C and thermal processing. *Innovative Food Science & Emerging Technologies, 23*, 107–113.

Santhirasegaram, V., Razali, Z., George, D. S., & Somasundram, C. (2015). Comparison of UV-C treatment and thermal pasteurization on quality of Chokanan mango (*Mangifera indica* L.) juice. *Food and Bioproducts Processing, 94*, 313–321.

Santo, D., Graça, A., Nunes, C., & Quintas, C. (2016). Survival and growth of Cronobacter sakazakii on fresh-cut fruit and the effect of UV-C illumination and electrolyzed water in the reduction of its population. *International Journal of Food Microbiology, 231*, 10–15.

Scheidegger, D., Pecora, R. P., Radici, P. M., & Kivatinitz, S. C. (2010). Protein oxidative changes in whole and skim milk after ultraviolet or fluorescent light exposure. *Journal of Dairy Science, 93*(11), 5101–5109.

Schenk, M., Loredo, A. G., Raffellini, S., Alzamora, S. M., & Guerrero, S. (2012). The effect of UV-C in combination with H_2O_2 treatments on microbial response and quality parameters of fresh cut pear discs. *International Journal of Food Science & Technology, 47*(9), 1842–1851.

Severo, J., de Oliveira, I. R., Tiecher, A., Chaves, F. C., & Rombaldi, C. V. (2015). Postharvest UV-C treatment increases bioactive, ester volatile compounds and a putative allergenic protein in strawberry. *LWT - Food Science and Technology, 64*(2), 685–692.

Shen, X. Z., Diao, E. J., Zhang, Z., Ji, N., Ma, W. W., & Dong, H. Z. (2014). Effects of UV-irradiation detoxification in a photodegradation reactor on quality of peanut oil. *International Food Research Journal, 21*, 2311–2314.

Shen, Y., Sun, Y., Qiao, L., Chen, J., Liu, D., & Ye, X. (2013). Effect of UV-C treatments on phenolic compounds and antioxidant capacity of minimally processed Satsuma mandarin during refrigerated storage. *Postharvest Biology and Technology, 76*, 50–57.

Sommers, C. H., Geveke, D. J., Pulsfus, S., & Lemmenes, B. (2009). Inactivation of *Listeria innocua* on frankfurters by ultraviolet light and flash pasteurization. *Journal of Food Science, 74*(3), M138–M141.

Sommers, C. H., Scullen, O. J., & Sites, J. E. (2010). Inactivation of foodborne pathogens on frankfurters using ultraviolet light and grass antimicrobials. *Journal of Food Safety, 30*(3), 666–678.

Spikes, J. D. (1981). Photodegradation of foods and beverages. In K. C. Smith (Ed.), *Photochemical and photobiological reviews* (Vol. 6, pp. 39–85). Boston, MA: Springer.

Stoops, J., Jansen, M., Claes, J., & Van Campenhout, L. (2013). Decontamination of powdery and granular foods using Continuous Wave UV radiation in a dynamic process. *Journal of Food Engineering, 119*(2), 254–259.

Tammineedi, C. V. R. K., Choudhary, R., Perez-Alvarado, G. C., & Watson, D. G. (2013). Determining the effect of UV-C, high intensity ultrasound and nonthermal atmospheric plasma treatments on reducing the allergenicity of α-casein and whey proteins. *LWT - Food Science and Technology, 54*(1), 35–41.

Taze, B. H., Unluturk, S., Buzrul, S., & Alpas, H. (2015). The impact of UV-C irradiation on spoilage microorganisms and colour of orange juice. *Journal of Food Science and Technology, 52*(2), 1000–1007.

Tran, M. T. T., & Farid, M. (2004). Ultraviolet treatment of orange juice. *Innovative Food Science & Emerging Technologies, 5*(4), 495–502.

Tripathi, S., & Mishra, H. N. (2010). Enzymatic coupled with UV degradation of aflatoxin B-1 in red chili powder. *Journal of Food Quality, 33*(s1), 186–203.

USFDA. (2000). *Ultraviolet radiation for processing and treatment of food* (No. 21CFR201.327, section 179.39 and 179.41). US Food and Drug Administration.

USFDA. (2001). *Hazard Analysis and Critical Control Point (HACCP): Procedures for the safe and sanitary processing and importing of juice: Final Rule* (No. Federal register 66) (pp. 6138–6202). Food and Drug Administration.

Vunnam, R., Hussain, A., Nair, G., Bandla, R., Gariepy, Y., Donnelly, D. J., … Raghavan, G. S. V. (2014). Physico-chemical changes in tomato with modified atmosphere storage and UV treatment. *Journal of Food Science and Technology, 51*(9), 2106–2112.

Wu, C.-K., Tsai, J.-S., & Sung, W.-C. (2015). Impact of ultraviolet treatment on improving gel strength of tilapia skin gelatin. *International Journal of Food Properties, 18*(8), 1702–1706.

Yang, S., Sadekuzzaman, M., & Ha, S.-D. (2017). Reduction of *Listeria monocytogenes* on chicken breasts by combined treatment with UV-C light and bacteriophage ListShield. *LWT, 86*, 193–200.

Yang, W. W., Chung, S.-Y., Ajayi, O., Krishnamurthy, K., Konan, K., & Goodrich-Schneider, R. (2010). Use of pulsed ultraviolet light to reduce the allergenic potency of soybean extracts. *International Journal of Food Engineering, 6*(3). https://doi.org/10.2202/1556-3758.1876

Zhao, X., Yang, W., Chung, S.-Y., Sims, C. A., Otwell, S. W., & Rababah, T. M. (2014). Reduction of IgE immunoreactivity of whole peanut (*Arachis hypogaea* l.) after pulsed light illumination. *Food and Bioprocess Technology, 7*(9), 2637–2645.

Zhu, Y., Koutchma, T., Warriner, K., & Zhou, T. (2014). Reduction of patulin in apple juice products by UV light of different wavelengths in the UVC range. *Journal of Food Protection, 77*(6), 963–971.

Index

S. Roohinejad et al. (eds.), *Effect of Emerging Processing Methods on the Food Quality*,
https://doi.org/10.1007/978-3-030-18191-8

MIX
Papier aus verantwortungsvollen Quellen
Paper from responsible sources
FSC® C105338

If you have any concerns about our products,
you can contact us on
ProductSafety@springernature.com

In case Publisher is established outside the EU,
the EU authorized representative is:
Springer Nature Customer Service Center GmbH
Europaplatz 3, 69115 Heidelberg, Germany

Printed by Libri Plureos GmbH
in Hamburg, Germany